T0281112

Cambridge Elements ≡

Elements in the Philosophy of Biology
edited by
Grant Ramsey
KU Leuven
Michael Ruse
Florida State University

ADAPTATION

Elisabeth A. Lloyd
Indiana University

CAMBRIDGE
UNIVERSITY PRESS

CAMBRIDGE
UNIVERSITY PRESS

University Printing House, Cambridge CB2 8BS, United Kingdom

One Liberty Plaza, 20th Floor, New York, NY 10006, USA

477 Williamstown Road, Port Melbourne, VIC 3207, Australia

314–321, 3rd Floor, Plot 3, Splendor Forum, Jasola District Centre, New Delhi – 110025, India

79 Anson Road, #06–04/06, Singapore 079906

Cambridge University Press is part of the University of Cambridge.

It furthers the University's mission by disseminating knowledge in the pursuit of education, learning, and research at the highest international levels of excellence.

www.cambridge.org
Information on this title: www.cambridge.org/9781108727549
DOI: 10.1017/9781108634953

First published 2021

A catalogue record for this publication is available from the British Library.

ISBN 978-1-108-72754-9 Paperback
ISSN 2515-1126 (online)
ISSN 2515-1118 (print)

Adaptation

Elements in the Philosophy of Biology

DOI: 10.1017/ 9781108634953
First published online: June 2021

Elisabeth A. Lloyd
Indiana University

Abstract: Natural selection causes adaptation, the fit between an organism and its environment. For example, the white and grey coloration of snowy owls living and breeding around the Arctic Circle provides camouflage from both predators and prey. In this Element, we explore a variety of such outcomes of the evolutionary process, including both adaptations and alternatives to adaptations, such as nonadaptive traits inherited from ancestors. We also explore how the concept of adaptation is used in evolutionary psychology and in animal behavior, and the adequacy of methods used to confirm evolutionary accounts of human traits and behaviors.

Keywords: evolutionary adaptation, evolutionary causes, development and evolution, evolutionary by-products, methodological adaptationism

ISBNs: 9781108727549 (PB), 9781108634953 (OC)
ISSNs: 2515-1126 (online), 2515-1118 (print)

Contents

1 What Are Adaptations?

1.1 Adaptation at the Organismic Level

Picture a woodpecker hanging on a tree. What do you see them doing? They pound their beaks into the bark of the tree at a rate of up to twenty times per second, so hard that the sound carries clearly throughout the neighborhood.

Woodpeckers have a number of special adaptations to aid them in getting the food they eat, which is grubs, insects, and worms that live underneath the bark of trees. What special adaptations help the woodpecker get at these insect foods they like to eat? And how do they escape the expected harm of slamming their heads and beaks up against a hard tree at a rapid rate for tens of minutes at a time? Wouldn't this scramble their brains, damage their eyes, and cause brain injury? Moreover, how do they reach inside the holes once they are made? Their closest relatives, the honeyguides, do not have extended tongues, but they do eat grubs and even beeswax from beehives. Let us consider these adaptive challenges met by the woodpecker.

First, the woodpecker has evolved a chisel-like tip on its beak to help drill holes in the trees in which its food lives. But wouldn't this repeated trauma to the head of the woodpecker damage the brain during this harsh drilling routine? No, because the woodpecker has also evolved a spongy lining to its skull, which has, in addition, thickened to absorb the drilling vibrations.

Moreover, the eyes of the woodpecker have also evolved special membranes to protect them during the harsh, repeated drilling motion. Finally, the woodpecker's tongue has evolved a barb at the end, and sticky saliva so that it can gather the insects inside the wood and bark of the trees after it drills its holes.

And there are additional special adaptations possessed by woodpeckers (also used by hummingbirds for identical purposes): The hyoid apparatus is lengthened. The hyoid apparatus is made up of a series of bones, muscles, and cartilage, connected to the tongue, that allow the tongue to extend to great lengths in nearly all of the woodpecker species. The hyoid apparatus wraps around the base of the skull and over and around the top of the skull, making the "base" of the tongue wrap fully around the skull of the woodpecker. In the red-bellied woodpecker, the tongue extends three times the length of its bill. We can see that the woodpecker species are specially adapted for their ecological niche of eating grubs and insects from trees with a suite of specially adapted traits enabling them to exploit this otherwise inaccessible source of food.

1.2 Evolutionary Models

Our scientific challenge is to explain how the woodpecker evolved all these special adaptations for its life of eating bugs from under the bark of trees. Any study of adaptation in evolution must rely on a very clear understanding of the process of natural selection, and also what kinds of evidence are desirable for demonstrating that a trait is an adaptation. We will review the basics of these selective explanations in this section, as well as the kinds of evidence that support such explanations.

What happens during a simple process of natural selection, according to evolutionary biologists? In all cases, we start with a population of entities, whether they are plants or animals, bacteria within our guts or groups of flour beetles, and variation in traits within that population.

Detailed aspects of the environment then constrain the reproductive success of some of those entities (call them the losing ones), while allowing or encouraging the reproduction of others among the population (call them the winning ones), depending upon their key features that vary. Under conditions where the key features of the winning and losing entities are reliably passed on to the next generation during reproduction, evolution by natural selection will occur within the population. Specifically, the proportion of entities with certain key winning features will grow, while the proportion of entities with the losing key features will shrink, over time (Futuyma & Kirkpatrick 2017; Griffiths et al. 2005). This embodies a simple process of natural selection, though not its only manifestation.

It turns out that selection does not act on traits separately, but rather as clusters or combinations. For example, in the woodpecker, it does no good to have an extra-long tongue if you do not also have the spongy protective skull to protect your brain from repetitive drilling.

One terminological point that we need to get to right away is: What exactly do we mean when we call something an "adaptation"? Do we mean that *any* trait that comes out of a selection process is an adaptation? Or do we mean that only traits that are "engineered" or whose form is changed by selection are adaptations?

Consider some examples to make the meaning of "adaptation" clearer. Many have learned about the case of "industrial melanism" involving the peppered moths in England. The population of moths was exposed to tree trunks blackened by the soot from factories, and the population of peppered moths went from mostly white moths to mostly black moths. This was due to natural selection on variation in traits in the form of birds eating the visible white moths that were resting on the black trees. Thus, the newly black dominance in the population can be considered an adaptation as a result of the selection process. But note that there are no new mechanisms and no new forms of moth in this case: There is no engineering or cumulative adaptation arising from the bird predation. Giorgio Airoldi calls this a case of "pure selection" (2018). All we have is a change in population distribution of the black and white moths. It is surely correct to call this a process of natural selection, and under the "selection-product" meaning of "adaptation," also correct to say that the black-dominance in the population is an adaptation. This "product of selection" (or "selection-product") definition views any outcome of a process that affects fitness in a positive manner as an adaptation. But it seems incorrect to say that there is a cumulative or engineering adaptation arising from this selection process.

Note the contrast with the woodpecker case. In the woodpeckers, we got a whole list of engineering adaptations that resulted, we believe, from natural selection processes and unknown amounts of developmental factors producing additional variation for selection to act on in these ancestral species of insectivores (species that eat insects). This "engineering" definition of adaptation has been dominant in the literature for many years and is the definition used in George C. Williams' ground-breaking book, *Adaptation and Natural Selection* (1966). Notably, Richard Lewontin (1978) and Stephen Jay Gould,[1] well-known critics of adaptationist research, also used the "engineering" definition of adaptation. The key relevant feature of an engineering adaptation is that it involves a mechanism or complex feature of some kind, built up cumulatively from simpler features, whereas the product of selection or distributive definition of adaptation involves simply a change or shift in allele frequency, one that could be unrelated to building a revised mechanism (Williams 1966, Lewontin 1978, Gould and Vrba 1982).

The "engineering" account requires that the adaptive account explain the adaptive feature(s) or modification(s) acquired through the selective process over time, its "engineering" history of its complex and built-up adaptive traits. Thus, the idea of an evolutionary *function* is tied intimately to the definition of adaptation itself. This is in accordance, for example, with John Maynard Smith's idea that "the 'function' of an organ is taken to mean those of its effects which have been responsible for its evolution by natural selection" (1978, p. 23). This could apply either to selection-product or engineering views of adaptation. But contrary to the usage of some other philosophers, I will be using exclusively the engineering definitions of adaptation and function throughout this Element, unless otherwise noted. See Table 1.1 for a more complete set of definitions below.

Later in this section we will talk about the notion of "function," but first let us consider some examples.

---

[1] Gould and Lewontin 1979.

Table 1.1 Definitions[2]

Term	Definition
Aptation	a trait that increases fitness (aptation for x – by x-ing). (If a trait is an aptation but not an adaptation, then it was not selected in the past for x-ing, and is an exaptation for x-ing).
Non-Aptation	a trait that does not increase fitness (but may do so in future)
"Engineering" Adaptation	a trait that involves a mechanism or complex feature of some kind, built up cumulatively from simpler features through natural selection processes. These types of adaptation involve both the etiological and the systems analysis types of 'function' (Williams 1966, Lewontin 1978, Gould and Vrba 1982; Lloyd 1988)
"Product of Selection" Adaptation	a trait that involves simply a change or shift in allele frequency, one that could be unrelated to building a revised mechanism (Williams 1966, Lewontin 1978, Gould and Vrba 1982)
Exaptation	a trait with no direct engineering function for x, which nevertheless increases fitness by x-ing (Gould and Vrba 1982)
Function	a trait has the engineering (evolutionary) function of x-ing, if x-ing increased fitness in evolutionary history, and the increased fitness explains the prevalence of the x-ing complex or engineering trait
Secondary Adaptation	a trait modified by natural selection (for x-ing, say) because of its contribution to fitness (by x-ing), where the trait so-modified existed, before modification, for a different reason than the role it came to have in x-ing
Spandrel	a trait is a "spandrel" if the trait in question (the trait that was pressed into service) for x-ing (where x-ing increases fitness but was not selected for its fitness contributions and hence does not have the function of x-ing) has no direct engineering function at all (a subset of exaptation).

---

[2] Thanks to an anonymous reviewer for help with these definitions.

Let's imagine, as Darwin did, a pack of wolves, some of which were swifter and slimmer than others in the pack: "[T]he swiftest and slimmest wolves would have the best chance of surviving, and so be preserved or selected" (Darwin p. 90, 1859).[3] They would do better in the evolutionary long run, reproducing more often and having healthier pups, thereby contributing more genes for the structure of their legs for swifter running to future generations through the process of natural selection. That's how the wolves became such swift runners over evolutionary time, able to take down very swift prey. In other words, we are claiming that "swiftness" is an evolutionary engineering set of adaptations in wolves, evolved by natural selection over evolutionary time. We have a type of biological model, in the claim of the process of evolution by natural selection, one that we can sketch and consider for its evidential weight. Or, we can ask, more specifically, what kinds of evidence would we need to establish a trait like swiftness as an evolutionary adaptation in wolves?

Now that we've had a chance to consider some adaptations in nature, we can see the various types of evidence needed to establish a trait as an adaptation. As Darwin argued, and as was later elaborated after the discovery of genetics, the key ingredients of the most basic natural selection model type in evolution are represented in Box 1.1.

We start with a population of organisms (and we fill in that blank in the selection model outline by specifying which population we are considering), and we have descriptions of traits to focus on, as well as claims of how these traits are heritable or based in genetics (filling in the appropriate blanks of the selection model outline). We also need claims about how these traits are related to fitness, usually supplied in the form of a mechanism explaining how the trait

---

Box 1.1 Natural Selection Model Outline

population [____]
variation in trait(s) [____]
genetic/cellular basis [____]
connection or mechanism between trait(s) and fitness [____]
selection pressure or environment [____]

---

[3] "[T]ake the case of a wolf, which preys on various animals, securing some by craft, some by strength, and some by fleetness; and let us suppose that the fleetest prey, a deer for instance, had from any change in the country, increased in numbers, or that other prey had decreased in numbers, during that season of the year when the wolf is hardest pressed for food. I can under such circumstances see no reason to doubt that the swiftest and slimmest wolves would have the best chance of surviving, and so be preserved or selected . . . . I can see no more reason to doubt this, than that man can improve the fleetness of his greyhounds by careful and methodical selection" (pp. 90–91, 1859).

increases fitness, as well as a description of a selection scenario that describes how pressure from an environment imposes on a population in a way that leads to changes in trait frequencies over time, as we saw with the fleet wolves and the woodpeckers.

As we can see, a variety of types of evidence might be required to substantiate the claims filling in the blanks of the selection model. Each of these claims, such as how the trait is related to fitness, and what the selection scenario is, needs to be substantiated with empirical or observational evidence.

Evolutionists Barry Sinervo and Alexandra Basolo (1996) offer a helpful discussion of what they think of as important evidence for adaptations. They start with a consideration of whether a particular trait of an organism is "optimal." Whether a trait is "optimal" is a calculation that depends on a separate model of *optimality*. Each optimal trait is the best-engineered trait it could be, given the constraints of the system, including developmental constraints and genetic constraints, and the job it needs to do, or the function it needs to fulfil.

Perhaps we can understand optimality better by looking at one of its most famous examples: dung fly mating times. Because males of this species of yellow dung fly mate with more than one female, they like to spend as little time as possible with each female, maximizing their chances at multiple matings with other females on other cow pies. However, the males also need to copulate with the females long enough to displace previous males' sperm inside the female's reproductive tract. The ability to do this depends on the size of the male. It is in the first male's interest to stick around after mating and guard her against mating soon with another male. Geoff Parker and colleagues (Charnov and Parker 1995) predicted that how long the male would copulate would depend on how far away the other cow pies were – in other words, how far away the other females were – and on how large the male was. This is an optimality model, in fact, a physical/biological version of an economics model called "marginal value theorem," and they were assuming that the male would optimize its behavior in terms of mating efficiency. And lo and behold, when they experimentally manipulated the dung fly males and the mating opportunities, they found that the males copulated with the females for the predicted periods of time (Charnov and Parker 1995).

Optimality models obviously make a number of assumptions about the systems they model, not least that the traits they model are engineering adaptations, but these models can be fruitful for research.

If we do assume that these traits are engineering adaptations, we may still want to know what the state of the traits was *before* the evolution of the adaptation in question. What did the species look like before the trait evolved the way it did?

Understanding this involves reconstructing the ancestors of the current population and trying to see what traits they had, without their relying on the adaptation we are studying. This may involve looking back into the ancestral tree of the population and seeing who is closely and more distantly related to the population, to see who had related traits and who did not. By doing this *comparative* work in *phylogenetics* (the study of the relatedness and traits of entities like species and lineages), we can better see what may be new in this species, lineage, or population of entities.

When studying the evolution of an adaptation, it may be ideal to do some experiments, as Sinervo and Basolo do (1996). The best experiments on adaptations often do two things at the same time:

(1) They manipulate key features of the environment, the very features that the trait is believed to be an adaptation for dealing with.
(2) And they manipulate the phenotype or appearance of the traits of the species being studied. So, for example, if we are studying the adaptation of the size of organisms, we would want to do an experiment that involves populations of both small- and large-size organisms.

When these experiments are done, it may be possible to determine which phenotype it is best to have in which environment. In other words, an organism may have higher fitness or reproductive success when it is large and living in large ponds, while a smaller organism may have higher fitness or reproductive success when it is living in smaller ponds. Thus, by manipulating both the environment and the phenotype, we are able to determine how the *fitness parameter* varies with the *trait* in a given environment. This is just the information we need in order to fill in the selection model.

Knowing all this, we might be able to discern whether the trait is an adaptation, that is, whether it evolved to serve a particular *function* in that species, given a certain environment. John Maynard Smith's idea is that "the 'function' of an organ is taken to mean those of its effects which have been responsible for its evolution by natural selection" (1978, p. 23).

In his review of the concept, Colin Allen offers us two main philosophical accounts of function:

> *Etiological* approaches to function look to a causal-historical process of selection; functions are identified with those past effects that explain the current presence of a thing by means of a historical selection process (typically natural selection in the case of biological function).
> *Systems-analysis* approaches invoke an ahistorical, engineering style of analysis of a complex system into its components. Functions of components

are identified with their causal contributions to broader capacities of the system. (emphasis added; Allen 2002, p. 375)[4]

Note that both the engineering adaptations and product of selection or effects type of adaptation, introduced above and present in Table 1.1, fall under etiological approaches to function; they both invoke the historical selection process to explain the current presence of a trait. The engineering adaptations also appeal to design analyses often identified with the systems-analysis notion of function. Thus, engineering adaptations in evolution appeal to both etiological and systems-analysis approaches to function in the philosophical senses just presented (Lloyd 1988; Williams 1966; Lewontin 1978; Gould 2002). Knowing all this, in addition, we may also be able to determine whether the trait is being maintained in current populations through current selection.

Adaptation is an "onerous concept," according to George C. Williams in his 1966 foundational text on the notion. He thought that a burden of proof rested on those claiming an adaptation and that adaptation should not be *assumed* to exist at the outset of biological investigation, just because a benefit could be perceived. We can look at how this burden of proof plays out by considering evidence supporting claims of engineering adaptations in guppies.

1.2 Confirming Evolutionary Models

1.2.1 Model Fit

There are several basic types of supporting evidence and confirmation:

The first type of evidence that can support the type of natural selection model we have just described is "model fit," that is, where the predictions of, for example, an optimality model "fit" or predict the **outcome** that we find in the real population of organisms. We can have model fit of selective models as well, where, after filling in all the blanks in our natural selection model outline in Box 1.1, we can predict correctly what the model outcome would be. We need an example to help understand this.

Biologists David Reznick, John Endler, and colleagues studied life history traits in guppies (see Figure 1.1). "Life history" traits are simply those traits involving the main stages in life, such as development to maturity, reproduction, and aging (Reznick and Travis 1996; Endler 1978).

---

[4] This latter systems definition is most often identified as a "Cummins function" approach (Cummins 1975). See Larry Wright (1973) for more on the etiological approach.

Figure 1.1 Guppies like those studied by Sinervo and Basolo (1996)

They studied the hypothesis that life history traits in the guppies were actually *adaptations to stage-specific predation* that were caused by predators, in other words, adaptations to the rate at which the predators of guppies killed the guppies. Reznick and his colleagues studied two communities that differed in their predators. One type had cichlids as predators – fish that prey heavily on adult-stage guppies – while the other had Rivulus fish that preyed only lightly on guppies, and when they did, it was mostly on juvenile guppies.

Evolutionary theories concerning predation and trade-offs between early and later reproduction made it possible to make predictions about life histories.

The evolutionists investigated the correlation between guppy life histories and the type of predator communities. The first prediction was that an increase in predation on the adults, like that found with the cichlids, would select for a decrease in the age at sexual maturity.

As we can see from Figure 1.2, the measurements from the guppies caught in the wild revealed differences in guppy life histories that were consistent with the predictions. In other words, there was good "model fit." Guppies from Rivulus localities, where predation was low, were older at maturity than their counterparts from Cichlid localities, where predation was high, which gave the guppies less chance to mature and breed at older ages. Thus, natural selection, according to the hypothesis, evolved guppies that matured earlier and were able to breed at younger ages in the high-predation environment, just as the model predicted. This is an example of directional selection, as illustrated in Figure 1.3. Simple directional or stabilizing selection usually produces *peaks* in the distribution curves of a trait, and with directional selection, the population average moves

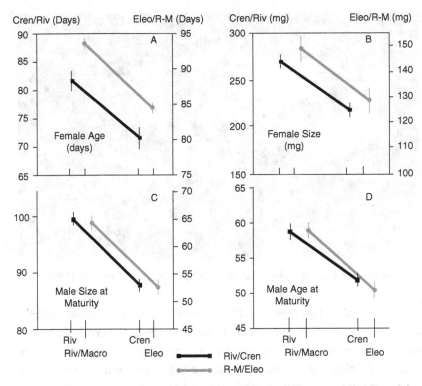

Figure 1.2 Decrease of age at sexual maturity depending on predator (Reznick and Travis 1996, p. 266): Least square means (+1 standard error) for the age and size at maturity and size at first birth in females from high- versus low-predation environments. See caption in Reznick and Travis for more information. Think of the predation rates as going from low to high from left to right on the *x*-axis. The stippled lines represent the means for the Low (Riv/Macro) versus high (Eleo) predation sites. The solid lines are the corresponding values for the high (Cren) and low (Riv) predation sites. All differences between the high- and low-predation sites were significant. (A) Female age at first birth (days). (B) Female size at first birth (wet weight in mg). (C) Male size at maturity (wet weight in mg). (D) Male age at maturity (days).

over to a new, more desirable, value. (We will examine a case of stabilizing selection in Section 4). Note that this account assumes that the traits are genetically controlled rather than arising from phenotypic plasticity.

This guppy reproduction case is a good example of the *good fit* of a selective evolutionary model prediction with the data from animals taken from the wild.

(a) Stabilizing selection

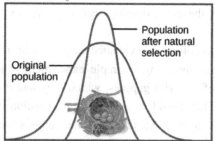

Population after natural selection

Original population

Robins typically lay four eggs, an example of stabilizing selection. Larger clutches may result in malnourished chicks, while smaller clutches may result in no viable offspring.

(b) Directional selection

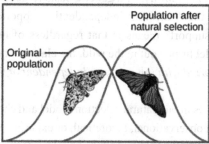

Population after natural selection

Original population

Light-colored peppered moths are better camouflaged against a pristine environment; likewise, dark-colored peppered moths are better camouflaged against a sooty environment. Thus, as the Industrial Revolution progressed in nineteenth-century England, the color of the moth population shifted from light to dark, an example of directional selection.

(c) Diversifying selection

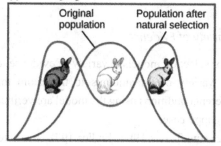

Original population

Population after natural selection

In a hyphothetical population, gray and Himalayan (gray and white) rabbits are better able to blend with a rocky environment than white rabbits, resulting in diversifying selection.

Figure 1.3 Selection regimes: stabilizing vs. directional selection[5]

1.2.2 Independent Support for Aspects of the Model

As exciting as it is to have selective model predictions be supported by evidence from the wild, there is another distinct type of evidence that can support selection models that may, in many cases, be even more important.

I call it 'independent support for aspects of the model.'

For example, any selection model includes the assumption that the trait under selection is "heritable," or reliably passed from generation to generation. This

---

[5] Image was taken from here: https://courses.lumenlearning.com/boundless-biology/chapter/adaptive-evolution/ License: *CC BY-SA: Attribution-ShareAlike Link to License:* https://creativecommons.org/licenses/by-sa/4.0/legalcode

usually means that the trait under selection has a "genetic basis," or, more accurately, some representation in the cells, development, or genes of the entities under selection.

In the guppy case, Reznick and his colleagues evaluated the genetic basis of the differences in life history traits by doing a simple experiment: They examined the second generation of the wild guppies, which they had bred and grown up in the laboratory. They found that the traits in question did reappear in this later generation, despite the fact that the guppies were removed from their standard selective environment and grew up in a simple glass tank. Thus, the assumption of heritability, or a genetic basis (a genetic/developmental/cellular basis) for the trait, was independently supported. What this claim of "independent support" means is that regardless of how the *predictions* of the selection model turned out in the wild, the claim that the trait had a heritable basis *was supported with lab evidence independent of that prediction.*

We could take several of the features in our natural selection model and check to see whether they had independent observational, theoretical, or experimental evidence to support them. These would all count as "independent evidence for aspects of the model."

1.2.3 Variety of Evidence

A third distinct type of evidence for selection models is variety of evidence, of which there are two kinds. The Variety of Fit amounts to the claim that *predictions or outcomes* under different conditions using the model are verified. We can see how this works in the guppy case.

Reznick and colleagues (Reznick and Travis 1996; Endler 1978) predicted that the guppies from the high-predation sites would devote a higher percentage of consumed resources to reproduction. The model predictions were confirmed in the first experiment, but then, in addition, the model predictions were confirmed in *forty new locations*, thus adding many cases amounting to a variety of fit (Reznick and Travis 1996).

Variety of Fit

Model predictions confirmed in a variety of locations can bolster support for a hypothesis. In a further example, when Reznick and colleagues tested guppy life history evolutionary traits across six independently evolved lineages in six independent stream locations in Trinidad, they bolstered their conclusions about those life history traits (El-Sabaawi et al. 2012).

Box 1.2 Filled-in Natural Selection Model

Population [*guppy population*]
Trait [*age of maturity*]
Heritability/genetic basis [*heritability out to two generations*]
Connection or mechanism between trait and fitness

[*early maturation leads to more offspring*]
Selection pressure [*age-specific predation*]

When an evolutionary model demonstrates compatibility with a very broad set of evidence like this, it is seen as being more strongly confirmed.[6]

There is, though, another type of variety of evidence to be gained by any evolutionary model, different from a variety of fit: the variety of evidence consisting in an accumulation of examples of independent support for aspects of the model.

Variety of Independent Support for Aspects of the Model

Let us turn again to the general outline of a selective model. This time, the blanks are filled in for the guppies.

Every selection model requires not just a selected trait, but also a claim about how certain versions or values of that trait are associated with better fitness values. Thus, independent evidence for this correlation is desirable. In addition, if the model also has evidence for the existence and strength of the claimed selection pressure, this lends even more support to the model. Reznick and colleagues (Reznick and Travis 1996; Endler 1978) checked the mortality rates of the different populations of guppies and showed that predation was the chief selection pressure.

In addition, to show more directly that predation drives evolutionary selection and change, Reznick and colleagues took guppies from high-predation environments and moved them to low-predation environments, and found that the high-predation guppies evolved the traits of the other population, with delayed maturity and decreased reproductive effort. All of this provides persuasive evidence supporting the "adaptation to mortality model" through independently and directly supporting its basic assumptions and dynamics rather than relying solely on successful predictions, which could provide only indirect support for the model's assumptions. Such independent confirmation of aspects of the model often turns out to be the most significant evidence offered for selection models (Lloyd 1983; 1988/1994).

---

[6] Supported by a technical approach, the Bayesian approach to probability logic; Glymour 1980; Fitelson 2001.

> ### Box 1.3 Variety of Robust Evidence
>
> More predictions were also confirmed, supporting the causal core of the selection models.
>
> The cause, higher predation, was also correlated with:
>
> - More frequent reproduction
> - Larger litters
> - Smaller offspring.

Robustness: Variety of Robust Evidence

Finally, robustness of evidence also plays a key role in supporting these evolutionary explanations. Reznick and Travis (1996) predicted that the guppies from the high-predation sites would devote a higher percentage of consumed resources to reproduction. This is basically to compensate either for more of their offspring being eaten by predators, or for the parent guppy having a lower likelihood of living to breed again. But other predictions also arose from the theory of the predator-rich environment, including more frequent reproduction, larger litters, and smaller offspring (see Box 1.3).

Once again, predictions from the model were robustly confirmed by evidence from the wild. This is an example of distinctly *different* predictions, involving different traits, being made using selection theory, and the fit of those predictions with the observational evidence. This is an example of a variety of robust evidence confirming the model and its causal core (Lloyd 2015b): the higher predation at the root of selection in the populations. This sort of "model robustness" has proved significant not just in evolutionary biology but also in climate science and economics.[7]

1.2.4 Totality of Types of Evidence Together

The selective model explaining the origin or maintenance of any adaptation is thus potentially confirmed or supported in a number of different ways by different kinds of evidence (Box 1.4).

1.3 Summary

Evolutionary adaptation occurs only as a consequence of a natural selection process; however, selection processes can occur in evolution without producing engineering adaptations. The evidence needed to document an engineering

---

[7]　Lloyd 2015b; Levin 1966; Weisberg 2006; Woodward 2006; Li and O'Loughlin manuscript.

BOX 1.4 TYPES OF SUPPORTING EVIDENCE FOR SELECTION MODELS

1. Fit
2. Independent support for aspects of the model
3. Variety of evidence
   a. variety of fit
   b. variety of independent support

4. Robustness: Variety of robust evidence

adaptation properly in evolution includes not only evidence of a contribution to fitness of a trait, but also evidence of its heritability and of a past selection process producing the engineering adaptation. Variations of these requirements occur at a variety of biological levels of organization, from gene to lineage to combinations of lineages, thus producing adaptations at the full span of biological levels, as we shall see in Section 2. There are, in addition, a variety of other factors involved in evolutionary change, which we will also review in Section 2 before turning our full attention to more theoretical issues involving adaptation.[8]

In Section 3 we will explore how adaptations are researched in evolutionary sciences, using a special technique from my philosophy of science work called the Logic of Research Questions. In learning to apply the Logic of Research Questions technique, the reader will apply evolutionary theory in context and learn a variety of evolutionary approaches. In Section 4 we will explore a few case studies, looking at whether these specific traits should be considered evolutionary adaptations or not, and how to determine the answer to such a question. We will look at the evolution of female orgasm, the determination of salamander toe number, and the evolution of a special genetic mutation in fruit flies. We will also consider the role and characterization of exaptations, a complementary concept to adaptation. We will conclude with some reflections on the difficulties of evolutionary exploration. While the subject is challenging, it is also very exciting, and we have abundant tools to explore the animal, plant, and microbial world.

2 Evolutionary Factors

In this section we consider a large variety of effective factors involved in evolutionary change that you may not have heard of, including variations on natural selection such as sexual selection and multilevel or hierarchal selection. We will explore how the giraffe is built, how human skulls are like snake skulls,

---

[8] **Suggested Further Reading:** R. C. Lewontin 1978.

and how butterfly wings got their spots, as well as what difference an organism's group makes to its evolutionary destiny. While organismic natural selection may have played a role in these features, it often did not play the most important role, and that is what interests us in this section. We take these various evolutionary factors from both traditional evolutionary theory, such as population genetics, and newer developments in evolutionary biology, such as evolutionary developmental genetics and beyond.

Our motivation for learning about this variety of evolutionary factors is to establish a fuller spectrum of causes for evolutionary change that might appear as legitimate alternatives or supplements to simple natural selection in an evolutionary explanation of a trait. As reviewed in Section 1, adaptation explanations appeal to natural selection as the causal factor. There are, however, many other causal factors known to have an impact on evolutionary change, most of which can interact with simple natural selection. Knowing more about these other factors can help us get a better and more complete picture of how evolutionary change takes place and can provide a clearer view of how evolutionary researchers might use these methods. Because we will be concerned primarily with the methods of natural selection research, we need to know what the *alternatives to* and *variations of* simple organismic natural selection are. We will start with classical population genetics and some sources of variation from that field.

2.1 Mutation

Genetic mutation is one of the ultimate sources of variety from which natural selection "chooses" variants in producing adaptive evolutionary change. (This "choice" is a metaphor for the selectivity of the process.) These mutations can be big or small in both impact and size, and are well understood in both their dynamics and components (i.e., DNA and/or RNA).

Mutations vary in how much of the genome they affect. "Point" mutations, which affect only a single DNA locus or base pair, are the smallest kind; but they can have large effects, as when the hemoglobin gene is affected in sickle cell anemia, producing a major disease when occurring in both chromosomes in an individual. (Note that when this specific point mutation occurs in only one chromosome, the person with the mutation is protected against malaria, a clear adaptation to environments with malaria.) We must also note, though, that much of the genome in multicellular organisms does not code for any protein or gene product. However, even changes in noncoding regions of DNA or RNA can affect how genes are expressed, making much of the genome consequential and subject to natural selection and other evolutionary causes that we will discuss in this section.

There are also structural mutations, which affect multiple DNA base pairs, from just a few to billions of bases. According to evolutionists Douglas Futuyma and Mark Kirkpatrick, most structural mutations happen as errors when chromosomes are replicated, and they include deletions, insertions, duplications, inversions, reciprocal translocations, and fusions (2017). There is also whole-genome duplication, which happens more frequently in plants, many of which are useful in the garden for their toughness, even though they cannot interbreed.

In any case, all of these genetic mutations comprise the foundation of variation upon which selection and all other evolutionary causal processes discussed in this section rest.

2.2 Migration

Another primary way that populations increase their variation is through migration from other populations with mutations or variants. While one population may be highly variable through a long history of interbreeding or "outbreeding" with various other populations, another population may have been relatively isolated, and hence lack variability from lack of migration from other populations. These differences in *gene flow* between populations can affect patterns of evolutionary change.

Gene flow or migration between populations can be either passive or active. In many plants, it is passive in that pollen is carried by the wind or by pollinators to other areas of the world with other populations, as when dandelion seeds and their parachutes are carried by the wind. In animals, much gene flow or migration is active.

Futuyma and Kirkpatrick offer the case of the desert locust or grasshopper, which usually is solitary, but sometimes swarms when conditions become crowded and resources become scarce (2017). Under crowded conditions, the grasshoppers' gene expression changes their shape, hormonal traits, and behavior, and, famously, they swarm in masses of hundreds of millions, flying long distances in search of better food sources. These activities spread the grasshoppers' genes far beyond their own neighborhood, thus providing a dramatic instance of genetic migration.

2.3 Drift

Drift has been understood as an original source of population variation from the earliest work in population genetics, due to Sewall Wright's studies of small populations and their random variability based on random sampling from a larger source population (e.g., Wright 1931). Drift is a classic example of how a population can fail to evolve an adaptation. The basic idea is that evolution can result from chance events of survival, reproduction, and inheritance; the resulting process is called *genetic drift*.

With randomly sorted small populations in a species, the distribution of genes within the populations varies a great deal. Thus, when one or more of these small populations becomes extinct, the random distribution of genes into the populations can "drift," or change, due simply to the random distributions of genes in the extinct populations. This leads to nonselective change in the genetic populations, or genetic drift.

Futuyma and Kirkpatrick show how drift can work with the grove snail, *Cepaea nemoralis*, which is famous for its colorful and highly variable shells (2017). These snails live in pastures that are shared by cattle and sheep, who often step on the snails accidentally and randomly, independent of their coloration. Nevertheless, this random crushing can have consequences for the balance of colors in later generations of snails; the chance obliteration of some colors of snails randomly affects the distribution of color genes in future generations, making this a case of *random drift*.

2.4 Phyletic Constraint

Consider the role of phyletic history, or as it is usually called, "phyletic inertia," due to lack of change from past history. The giraffe's larynx or voice box is an intriguing case of how evolution does not often invent new things wholesale. In other words, evolution usually works as a tinkerer within constraints of the present design, a principle that is generally accepted and uncontroversial (Jacob 1977). Let's look at the giraffe's "design."

Controlling the larynx is the recurrent laryngeal nerve, and this is true from fish to mammals in evolution; the same nerve – called a homology (using the developmental biologists' conception of homology)[9] because it has the same origin in the genes and cells – appears in development, but it traces different pathways within the body of different animals because of their dissimilar development. Evolution never starts from scratch in developing or "designing" an organism. Remember, evolution is a tinkerer, changing this or that bit of an organism to make it more suitable or a better fit with its environment in this or that way, given the parts available to it, and given the developmental and genetic changes available at a given time to the population.

---

[9] There has been some controversy about the developmental biologists' conception of homology because the phylogenetic definition of homology, under which two traits are homologous if they descend from the same trait in a common ancestor, is more commonly used.

For example, in fish, the laryngeal nerve traces within the skull directly from the brain down to the larynx. Through evolutionary tinkering over time, because of how mammals evolved from fish and then reptiles, in human beings the nerve first loops down to the chest and around the aortal arch in the chest, and then back up to the larynx, which is the long way around. In evolutionary development, certain things can change, like the structure of the head, neck, and thorax, but others cannot, like the nervous system basics, apparently. In other words, selection acts on existing variation, and the sources of this variation depend on developmental processes and history.

In the giraffe, the looping of the laryngeal nerve reaches the peak of absurd shape and development, forcing the nerve to go from the head all the way down the neck to the chest cavity where the aortal arch is, and then all the way back up the neck to the larynx. This looping effect, the result of developmental change within constraints of the present design of the nerves and evolutionary tinkering over time, makes the nerve about fifteen feet long, and it is thought that this feature limits the range of vocalizations that giraffes are capable of (Harrison 1981). In

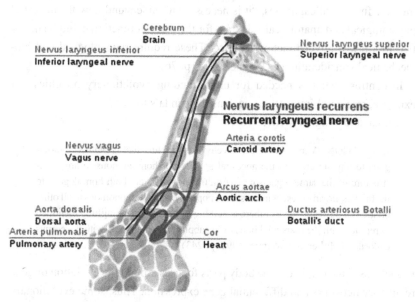

Figure 2.1 Scheme of path of the recurrent laryngeal nerve in *Giraffe camelopardalis*.[10]

---

[10] Wikimedia Commons: Scheme of path of the recurrent laryngeal nerve in *Giraffe camelopardalis*. Date: February 21, 2010. Created in Inkscape using *Oakland_Zoo_dsc_2658.jpg* and information from R. J. Berry and A. Hallam (eds.), *The Collins Encyclopedia of Animal Evolution* (London: Collins, 1986). Author: Dr. Bug (Vladimir V. Medeyko).

certain extinct dinosaurs such as *Supersaurus*, the laryngeal nerve may have been up to 100 feet long, and it is believed to have limited their vocalizations as well.

Thus, evolution as a tinkerer, working within the constraints imposed within a phylum (or large group of related species), can result in traits that are far from being ideally adaptive to their environment. This is an example of "phyletic constraint," in technical terms, within evolutionary biology, and it is a classic constraint within evolutionary theory on ideal adaptive engineering.

2.5 Developmental Constraints and the Origin of Evolutionary Novelty

It might be helpful here to distinguish between innovation and novelty, both of which are related in different ways to adaptation. Following the distinction promoted by biologist Günter Wagner (2015), we can think of evolutionary innovations as referring to novel functional capabilities, like flight or bipedal walking (Müller and Wagner, 1991; Love 2003, 2006), and evolutionary novelties as structural innovations such as the panda's thumb. With regard to explaining new functional capacities, "it is necessary to understand how the ancestral physiological and anatomical substrate for the derived function could be modified to allow a novel function to arise. These require physiological and biomechanical considerations" (Wagner 2015, p. 76).

In contrast, what is needed for understanding evolutionary novelties, for example, structural innovations such as the panda's thumb?

Muller and Wagner write:

> DEFINITION: A morphological novelty is a structure that is neither homologous to any structure in the ancestral species nor homonomous to any other structure of the same organism ... Additional bristles are both homologous to the bristles already present in the source population and homonomous to all other bristles on the same fly. But there is nothing that can be meaningfully identified in reptiles with the marsupial bone or in subplacental mammals with the corpus callosum. (Muller and Wagner, 1991, p. 243)

In such cases the origin of new body parts focuses around the evolution of gene regulatory networks and differential gene expression. Thus, some evolutionary explanations focus on the development or embryology of the organism. Darwin himself called attention to this type of evolutionary explanation in his discussion of the joints of the bones in the skulls – called sutures – in birds and reptiles.

In Figure 2.2, you can see these joints in the skull of a king snake on the left, and the similar joints or sutures in the skull of a human infant (right). In Darwin's time, the joints were thought to be a special adaptation for birth in

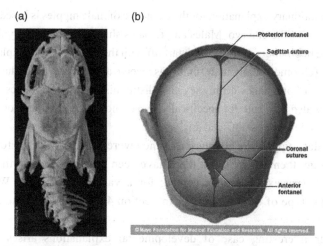

Figure 2.2 Snake skull [left] compared with human skull [right]: Note lines (sutures) in skull. These are places where the bones grow together to form the complete skull, in both snakes and human beings. Snake skull image credit: Digimorph, An NSF Digital Library at UT Austin. Human skull image credit: Mayo Foundation for Medical Education and Research.

mammals, since they made the skull more flexible and compressible to go through the birth canal.

But Darwin disagreed:

> The sutures in the skulls of young mammals have been advanced as a beautiful adaptation for aiding parturition [birth], and no doubt they facilitate or may be indispensable for this act. But, as sutures occur in the skulls of young birds and reptiles, which have only to escape from a broken egg, we may infer that this structure has arisen from the laws of growth, and has been taken advantage of in the parturition of the higher animals. (*On the Origin of Species*, p. 197)

In other words, the sutures in the skulls of young mammals may be essential for helping them squeeze through the birth canal, but this is not their reason for existence. Rather, the sutures came into existence because of the necessities of development, going all the way back to the reptiles, and the uses to which the sutures are later put by evolution are bonus side effects.

For a related evolutionary challenge, consider the problem of why male mammals have nipples. Nipples clearly provide a reproductive advantage to *female* mammals in that they ensure reproductive success by providing the means to feed the offspring. But there is no known contribution to fitness for the males.

The evolutionary explanation for the existence of male nipples is based on the development of the embryo. Males and females share the same embryological form at the beginnings of life – they start off with the same basic body plan, and only if the (chromosomally male) embryos receive a jolt of hormones during the eighth week of pregnancy do any sexually distinguishing characteristics appear. There may also be a role for specifically female hormones and developmental cues, but we haven't established any yet.

In females, nipples are adaptations – they were actively selected for – but the males get them for free. But there have been problems getting this kind of evolutionary explanation accepted, for a variety of reasons. We will discuss this type of example further in Section 4, in the case of the female orgasm.

Another interesting case of developmental explanations arises in the explanation of why so many miniaturized salamanders have only four toes instead of the usual five toes of land animals. There are numerous lineages in which the salamanders become miniaturized, or dwarfed, and in all of those lineages, the miniaturized forms converge on having only four toes on their rear legs instead of the five toes of the usual-sized salamanders and other land animals and four-legged creatures, such as you can see in Figure 2.3 of the independent lineages.

Why so? Is selection driving this toe reduction, so that all lineages are under selection when tiny to produce the four-toed form? So that there is a functional or adaptive advantage in every lineage to having only four toes in the available environments, and this selective pressure is expressed in every lineage? Or is there another, more form-based evolutionary explanation, where only limited developmental and structural options exist in the lineages, and once miniaturization happens, only the four-toed forms are really available? Actually, these explanations are complementary and are likely best combined, according to Wake.[11] This case is discussed further in Section 4.2.

Another important place for development is in the origin of developmental novelty. For example, in the case of the evolution of the eyespots on butterfly wings, which are relatively recent novelties or innovations on the wing coloration of butterflies, they are understood to be adaptive for avoiding predators through camouflage. In order to tell their full evolutionary story, we need evolutionary developmental biology to tell us where the novelty originated, which we can learn from the developmental genetics of butterflies and related organisms (Wagner 2000, p. 96).

---

[11] See discussion by Griesemer 2013, 2015.

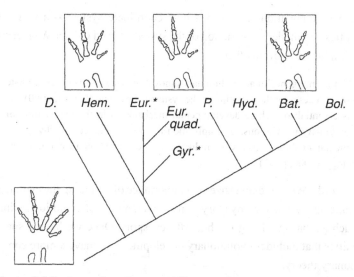

Figure 2.3 Independent four-toed lineages in the salamander family Plethodontidae. Independent evolution of four-toed state in three separate lineages. (Wake 1991, p. 548, Figure 1).

It turns out that eyespots are completely unique to butterflies, so they are genuine evolutionary novelties. The genetics show that they are induced genetically by a small group of cells called the "eyespot organizer," so the evolutionists concluded that the key evolutionary step occurred in the origin of this organizer (Nijhout 1991). Butterflies and fruitflies are from the same assemblage, Panorpida, within the superorder of insects, *Endopterygota* (Song et al. 2018). (The interested reader can check out a visual depiction of their relation on the web, for example, Wikipedia.)

So, the evolutionary developmental geneticists studied the genes in fruit fly wing development and how the same genes were involved in butterfly wing development. Except for butterfly wings, the genes had acquired a new function, an innovation. They then determined exactly the two genetic modification events that were required to result in the eyespot organizers. This knowledge about the evolutionary novelty requires detailed knowledge of both fruit fly and butterfly wing development. And the genetic architecture of butterfly pigment patterns "radically changes as a consequence of the origin of eyespots" (Wagner 2000, p. 97).

What this means is that any regular population genetics model of evolution by natural selection and other forces of these butterflies would not be effective, because the entire range of possibilities of the model would be incorrect; it would be transformed by the evolution of the novel eyespots. After the evolutionary innovation, we need a new coordinate system – a new set of possibilities – to model the variation of the system.[12] And Günter Wagner spells out the implications of this fact:

> after the emergence of an evolutionary innovation, we need a new coordinate system to describe the variation of the system (Shpak and Wagner, 2000) . . . .
> it is essential to include developmental mechanisms in the explanation of evolutionary innovations. By implication this is also the reason why developmental evolution makes an indispensable contribution to evolutionary biology. (2000, p. 97)

In other words, Wagner emphasizes the importance of including the evolution of developmental systems in any story of the development of evolutionary innovation such as that involving the butterfly eyespots. Once we have a space of possibilities that includes evolutionary development, we have a more complete evolutionary theory.

2.6 Genetic Linkage

Genetic linkage means that genes are being inherited together at a higher rate than random. They are another illustration of developmental constraint or forces of cellular structure and development making a difference to evolutionary change.

Because of genetic linkage, adaptation may be constrained, due to a lack of genetic variation for selection to act on. With less genetic variation for selection to act on, adaptation will have a more restricted array of adaptive traits from which to choose, thus limiting the level of adaptation in the organisms' population.

Human genes are lined up on two sets of chromosomes. When two alleles (versions of a particular gene) are close together on a given chromosome, they are more likely than not to be inherited together during the processes of reproduction, a phenomenon called "genetic linkage." During meiosis, the chromosomes cross over, and some material from the mother's chromosome mixes with some material from the father's chromosome. The further apart the alleles are on each chromosome, the more likely it is that they will be separated, and *not* linked, during meiosis and reproduction.

Most such genetically linked genetic traits are either closely spaced together on a single chromosome, or they govern a common pathway. We can discover which genes are linked by comparing genetically crossed plants' or animals' genes to calculate the recombination frequency, which measures how often the genes change position. Two genes that are very close together will have few recombination events and therefore a low recombination frequency.

---

[12] Shpak and Wagner 2000.

By finding recombination frequencies for the various gene pairs, we can create linkage maps that show the order and relative distances of the genes on the chromosomes. This is how the human genome was first mapped and its DNA was sequenced.

2.7 Pleiotropy

Pleiotropy occurs when single genes have multiple phenotypic effects or traits. The originator of genetics, Gregor Mendel, conducted pea studies that revealed pea plants with pleiotropy – genes that had more than one trait associated with them. For example, the plants with colored seed coats always had colored flowers and colored leaf axils (the part that attaches leaves to stems). The opposite was true of colorless seed coats, which had white flowers and no pigment on their axils. Mendel's findings were later confirmed to be the results of single genes for these multiple traits.[13]

We are familiar with pleiotropy in white cats with blue eyes, 40 percent of whom are also deaf. It turns out that pigmentation plays a role in maintaining fluid in ear canals, without which there is deafness (Sunquist, 2006). This is a common case of pleiotropy, and even works partway, in that a cat with one blue eye and one yellow eye is deaf only on the blue-eyed side of its head.

Thus, features or traits of genes can appear that are not separately selected for – they just come along for the ride. In the case of the peas, these may be benign traits, but in the case of deafness, they can be maladaptive, even though the white fur and blue eyes themselves may be prized by certain cat owners. The crucial point is that we cannot infer that every trait is an adaptation, selected by natural selection for its current state or value; it may be a ride-along trait of pleiotropic genes.

2.8 Cultural Coevolution Factors

For processes like cultural selection, our best example may be the gene-cultural coevolution of *lactose intolerance*, or the adult inability to digest milk (Feldman and Cavalli-Sforza, 1985). The basic idea is that culture can shape biological evolution. Dual inheritance theories of genes and cultural inheritance not only track two different types of inheritance but can also represent their dynamic feedbacks when it occurs.

Feldman and Cavalli-Sforza initiated gene-culture evolutionary models in 1973, when they introduced a model that incorporated cultural transmission (see also Feldman and Cavalli-Sforza 1981). The evolution of lactose absorption

---

[13] Fairbanks and Rytting 2001.

represents a good example of gene-culture coevolution. There is systematic variation in adult humans' ability to digest milk. In fact, most adult humans cannot digest it, and their lack of lactase, the enzyme that breaks down milk sugar, lactose, means that when they drink milk, they are likely to get stomachache, nausea, and diarrhea. There is a genetic basis to the differences between people in their ability to drink milk and eat milk products, based on a dominant genetic trait, and this genetic difference correlates to the history of dairy farming in the relevant populations and their cultures. In cultures with dairy farming, over 90 percent of the population's adults usually can digest milk, while in nondairy farming cultures, the rate is less than 20 percent. It is believed that, since dairy has been an important part of the diets of some human groups for over 6,000 years, such cultural practices may have set up a selective regime under which the allele for lactase selection was favored.

Feldman and Cavalli-Sforza (1985), following work by Aoki (1984;1986), used gene-culture coevolutionary theory to investigate the evolution of lactose absorption. They modeled lactose absorption with differential cultural transmission of milk usage. Their analysis found that whether or not the allele for milk absorption was favored depended crucially on the likelihood that the children of milk users become milk users themselves, in other words, whether they continued the cultural practices of their parents. Thus, the cultural practices and processes complicated the selection process, often to the extent that the outcome changed from the one expected under purely genetic results. Thus, gene-culture coevolutionary models can be very informative, and interesting results can explain a variety of outcomes in human society.[14]

2.9 Niche Construction

Lewontin (1983) proposed that biological evolution depends on "niche construction," as well as natural selection and inheritance. The notion of niche construction has been developed since then by Kevin Laland, John Odling-Smee, and Marcus Feldman into a theory about the activities, choices, and metabolic processes of organisms, "through which they define, choose, modify, and partly create their own niches" (2000, p. 132). We can see this when organisms choose their own habitats, their own mates, and then also choose resources and construct important parts of the local environments, for example, webs, nests, or burrows, but also chemical environments.

Organisms also partly destroy their own habitats by stripping them of resources or building up waste. Niche construction can interact with natural selection in interesting ways. For example, organisms can counteract natural

---

[14] Feldman and Laland 1996, p. 454. See also Boyd and Richerson 2009.

selection by digging burrows that protect them from the cold. On the other hand, they may provoke natural selection through niche selection choice of a new food source, subsequently selecting for a new digestive enzyme. Or they might do both, for example, when building a nest protects an organism but also establishes a novel selection pressure by acting on a second trait for defending the nest. In each case, though,

> [N]iche construction modifies one or more sources of natural selection in a population's environment and, in so doing, generates a form of feedback in evolution that is not yet fully appreciated. (Laland et al. 2000, p. 133)

This feedback generated through niche construction constitutes the primary evolutionary significance of niche construction. Many organisms modify the selection pressures on themselves through niche construction, and these selection-altering traits coevolve with traits whose fitness depends on changeable sources of selection in environments. For example, take the beavers' niche construction activity of building a dam. The dam itself sets up a variety of selection pressures on the beavers, and these pressures feed back to act on the genes responsible for the dam. Take the traits such as the details of the teeth that must cut down the trees to make the dam, the tail as a balancing and warning tool, the feeding behavior that is changed through having a dam, and the social system once the dam is set up – all of the genes for these traits are under different selective regimes once the dam exists. Also, the dam will affect many future generations of beavers that inherit the dam and the altered stream, which is now a pond, and a source of many sources of food and shelter. The feedbacks produced by the simple existence of the niche alteration of the dam are enormous for evolution.

And that's just one example. Think of ants who build their nests, complicated structures around which they move their broods of eggs, or think of bees, who build hives with similar complexity and defenses. There is strong evidence that the burrow or nest came first, and the subsequent defenses and elaborations evolved later, building on top of the initial move in niche construction.

2.10 Sexual Selection

Darwin (1859, 1871) proposed the mechanism of sexual selection to explain evolution of conspicuous traits in males that reduce their survival, such as bright plumage and vigorous courtship displays. Christopher Murphy defines sexual selection as "differential mating success produced by variation among members of one sex in a trait that affects competition for mates" (1998, p. 8).

Darwin defined two major types of sexual section:

> [I]n the one it is between the individuals of the same sex, generally the male sex,
> in order to drive away or kill their rivals, the females remaining passive; whilst
> in the other, the struggle is likewise between the individuals of the same sex, in
> order to excite or charm those of the opposite sex, generally the females, which
> no longer remain passive, but select the more agreeable partners. (1871, p. 398)

Subsequent discussions of sexual selection have focused on these two types of sexual selection, male–male competition and female (mate) choice. A number of theorists have suggested additional mechanisms, and Murphy has generalized that traits can be subjected to sexual selection "even if they do not affect the outcome" of selection between individuals of the same sex or different sexes, that is, they are totally independent (1998, p. 16).

He distinguishes two types of sexual selection based on whether variance in mating success is the result of "trait-dependent outcomes of interactions between" individuals of the same species (1998, p. 9). He uses this distinction to construct a framework for classifying different types of sexual selection previously proposed.

Lynda Delph and Tia-Lynn Ashman (2006), Steve Arnold (1994), and Delph and Christopher Herlihy (2012) distinguish selection for survival (or viability), for mating success (or sexual selection), and fecundity (or fertility selection), and how they act within each sex to determine the evolution of sexual dimorphism —or differences in form in the different sexes— in *Silene latifolia*, a wildflower common in North America. Delph and Herlihy experimented on *Silene* by varying distinct selection regimes that differed in flower size, a sexually selected trait. In *Silene*, flower size and number trade off within each sex, depending on the selection pressures. And males produce more numerous and smaller flowers than females.

The outcomes of the selection experiments that placed the different varieties of selection pressures on the plants were as follows. Fecundity selection favored males with high flower production, since it was correlated with pollen production, key to success in fecundity. Also, viability success opposed success in sexual selection, favoring males from the large-flower lineage. But in females, fecundity and viability selection operated in the same direction, favoring those with greater seed production and survival, which were also those from the large-flower lineage (Delph and Herlihy 2012). Thus, these three types of selection can interact in a variety of ways to produce evolutionary outcomes.

We also see sexual selection acting versus viability selection on size in the marine iguana in the Galapagos Islands in an unusual way, as they adapt to El Niño temperature changes in their environment, in a study by Martin Wikelski and Corinna Thom (2000).

Changes in bone metabolism allow these lizards to reversibly alter their length as adults in a mechanism unknown in other vertebrates. In two long-term studies of two iguana populations, one of 18 years and one of 8 years, the lizards shrank by as much as 20 percent when exposed to severe food shortages brought on by El Niño conditions. When La Niña conditions resumed, the alga food sources were renewed, and the adult iguanas resumed their previous lengths. The iguanas that shrank more survived longer than the larger iguanas and were able to feed more. Larger individuals shrank more than smaller ones, and females shrank more than males of the same size. This was believed to be because they exported their energy into clutches of eggs in the year before an El Niño.[15] This shrinkage, of up to 20 percent, is mostly due to bone absorption, caused by energetic stress and low food availability.

2.11 Multiple Levels of Selection

It is important to realize that evolutionary adaptations can occur at many levels of biologic organization, including the genic, cellular, organismic, family, demic, lineage, and mutualist/holobiont levels.

2.11.1 Genic and Cellular Levels

At the genic level, we have cases of *meiotic drive*, wherein sperm cells possess the means to kill off competing sperm cells in a manner that favors inheritance of their own genes, for example, in the t-allele in the house mouse (Lewontin and Dunn 1960). The t-allele causes sterility in the male mice it appears in, and demes or groups in which the t-allele is more common clearly have a disadvantage in reproduction. Still, the t-allele persists in the population because it outcompetes other sperm cells at the genic level within gametic reproduction, even though it is selected against at the two higher levels of selection.

On the cellular level, we have the cooptation of eukaryotic (nucleus-containing) cells by free-living bacterial mitochondria, wherein we understand the previously independently living prokaryotic (non–nucleus-containing) mitochondria have been coopted and absorbed cooperatively and symbiotically into the eukaryotic cell (Margulis [AKA Sagan] 1967)). The new eukaryotic cell reproduces both its own DNA and the DNA of the mitochondria in parallel.

2.11.2 Symbiosis Level

These mitochondria are early cases of holobionts, where two independent sets of organismic genomes are linked together through coadaptation and coevolutionary

---

[15] Wikelski, Carrillo, and Trillmich 1997.

change. Other examples occur with animals and their gut microbiomes (communities of microbes), such as cows and the bacteria, viruses, and fungi that they depend on to digest the cellulose in their diet; similarly for termites, who require their gut microbiome to digest the wood fibers in their diet. There are adaptations that can emerge at the coadaptive level, such as the pedestals built by dung beetles on which they deposit their eggs. When the eggs hatch, they are surrounded by the dung, which contains the gut microbiome that will allow them to thrive in their environments (Lloyd and Wade 2019; Suárez and Triviño 2020; but see Douglas and Werren 2016).

The building of the pedestals made of dung is a joint adaptation of the mutualistic holobiont itself, serving its shared, joint (epistatic) fitness component, as modeled by Lloyd and Wade (2019) and as theorized as a "manifester of adaptation" by Suárez and Triviño (2020; see Lloyd 2017).[16] While we model such a holobiont as a sort of community or group, let us now consider a more general group-level approach.

2.11.3 Group Level

Consider a group selection case from Michael Wade and community selection case from Charles Goodnight. In the group selection case we must consider genetic interaction, that is, the fact that individual genes' selective fitnesses are never considered additively in the system, but always in relation and interaction with the other genes in the system, including genes of other organisms in the same social group or family as the original organism.[17]

Wade offers two card games to help understand what is going on in group selection like this. Take the card game of War, wherein each player lays down a card at each play, and the highest card wins. The cards are totally independent, and there are no interactions with the other cards in the deck. In poker, however, the value of each card depends very much on what other cards you hold in your hand. "Interactions are the essence of poker . . . .[it has] many possible winning hands, depending on the combinations of cards, the spatial and temporal order in which hands are played, the abilities of the other players in the game, and the past history of hands played by the 'population' of players" (Wade 2016, p. 190).

There is much evidence for the effectiveness and power of group selection as defined by Wade and Goodnight, both in the laboratory and in the wild. Group

---

[16] This means that there may be a component of fitness of the mutualistic *combination* of organisms, or a *shared* epistatic component of fitness (Lloyd and Wade 2019).

[17] Wade 2016, p. 188.

selection can produce group adaptive and behavioral changes that allow the group to thrive in otherwise inhospitable conditions.

While many of those taking a genic approach are unaware of the power of group selection, this does not change the status of the empirical evidence nor the likely efficacy of group selection in producing evolutionary change in populations and species in the wild (see Lloyd 2005b)[18]. One of the most striking examples of the success and power of group selection is in the domestic chicken. Chicken breeders had long selected for even tiny improvements in egg-laying rates of hens under domestication, but had not measurably improved their rates due to conflict among the hens in captivity, and their high mortality rate. Group selection on the hens for hen groups that got along better and were more docile, the "kinder, gentler" hens, improved the egg-laying productivity of the hens by up to 60 percent (Cheng and Muir 2005), thus demonstrating the adaptive power of group selection regimes.

Sometimes when multiple levels of selection are understood to occur simultaneously, this is referred to as "hierarchical selection." For the classical modeling of such systems, see Sewall Wright (1931), or for its first real-world instantiation, see Lewontin and Dunn's work on the house mouse with its t-allele (1960).

2.11.4 Lineage Level

Lineage levels of selection may also be effective in producing evolutionary change, at the very least, under the description of product-of-selection adaptations rather than engineering adaptations (see Section 1). Consider, for example, the rates at which lineages speciate, that is, the rate at which they form and branch off new species. This rate varies among different lineages, and interestingly, it varies with the geographic distribution of the species, that is, how the lineage ranges over the Earth, and how it occupies different sectors or areas of the Earth's geography.

Elisabeth A. Lloyd and Stephen Jay Gould (1993) followed the line of thought that lineages of gastropods might develop higher speciation rates depending on their breeding mechanisms, and may also vary in their long-term species survival rates. Those lineages that reproduced in such a way that the young of the species were sedentary and attached themselves to rocks at the beginning of life, thus limiting their range very sharply, were seen as more likely to speciate. The

---

[18] Against a conventionalist interpretation of genic selectionism, à la Sterelny and Kitcher (1988) or Kerr and Godfrey-Smith (2002), see Lloyd (2005b) and Lloyd, Lewontin, and Feldman (2008), showing its undesirability and the nonequivalence of the claimed group and genic models.

lineages that reproduced by having the young float freely in the seawater were able to travel far from their place of birth and cover a greater geographical range, and were seen as less likely to speciate. Later, the lineage-level adaptive trait of geographic range of offspring in both gastropods and mollusks was found to be correlated with the long-term species survival rate (Jablonski and Hunt 2006; Jablonski 2008). This accorded well with the Lloyd and Gould theory of species selection, which does not require engineering adaptations at the species level, but finds product-of-selection adaptations nonetheless.

2.12 Evolutionary Mismatch

Evolutionary "mismatch" is a state of disequilibrium, "whereby a trait that evolved in one environment becomes maladaptive in another environment. Mismatch is an integral part of evolution in changing environments and is becoming increasingly common for all species living in human-altered environments" (Lloyd, Wilson, and Sober, 2011). In a classic case of mismatch, a trait that is adaptive in an ancestral environment becomes maladaptive in a modern environment, or ill-suited to perform its function. The situation can typically only be addressed by a behavioral accommodation, subsequent evolution, or another environmental change.

Take baby sea turtles, who have adapted to move toward the light of the moon reflecting off the sea as soon as they hatch on the beach, an adaptation that leads them immediately toward the ocean. Once beach houses, with their deck lights are built, however, the baby sea turtles become confused and disoriented, moving further inland toward the beach houses and away from the ocean, which is maladaptive, a case of evolutionary mismatch. The originally adaptive trait of moving toward light has now become maladaptive in this new, modern, beach-house environment. The solution for protecting the sea turtles is to build walls to contain the light within the boundaries of beach-house properties, thus allowing them to react to natural light and move once again toward the ocean.

Evidence for evolutionary mismatch or maladaptations follows the same route as that for adaptations, differing primarily in the extra step involving a link between fitness and the trait, and its environment of selection. In evolutionary mismatch or maladaptations, the new link requires evidence for a past link between fitness and the trait, along with evidence for a present negative link (e.g., linking the turtles' maladaptive behavior with the new beach-house environment). The literature about mismatch holds various views about whether an actual negative fitness value is required; some require only damage to well-being (e.g., cancer or diabetes onset that can occur after reproductive age).

But all the usual assumptions and steps in providing evidence for an adaptation need to be filled in. Thus, an adaptive explanation for a trait is not confirmed

---

Box 2.1 Evolutionary Mismatch Model and Confirming Evidence

population [_____]

trait(s) [_____]

genetic basis [_____]

past positive or neutral connection or mechanism between trait(s) and fitness [_____]

past selection pressure or environment [_____]

present negative connection or mechanism between trait(s) and fitness [_____]

present selection pressure or environment [_____]

---

primarily by providing a story or prediction that says that it would be good for an organism to possess the trait in question, but rather by a variety of detailed biological measures and assumptions. We can see how Box 2.1 would be filled in for the sea turtle case, with the present negative connection between trait and fitness being the death of baby sea turtles who climb up the beach toward the beach houses, and the present selection pressure being the presence of beach-house lights attracting the baby sea turtles.

2.13 Summary

Evolutionary adaptation occurs only as a consequence of a natural selection process; however, selection processes can fail to produce engineering adaptations. The evidence needed to document an engineering adaptation in evolution includes evidence not only of a contribution to fitness of a trait, but also of its heritability, and of a past selection process producing the engineering adaptation. Variations of these requirements occur at a variety of biological levels of organization, from gene to group to lineage to combinations of lineages, thus producing adaptations at the full span of biological levels. In this section we have given a whirlwind review of a variety of factors that play roles in evolutionary change over time.[19]

I have reviewed a variety of ways in which organisms can evolve, with a variety of selective and nonselective causes producing these outcomes. As I hope the reader can see by now, evolutionary research is complicated by the presence of this full variety of evolutionary factors, because any one or

---

[19] **Suggested Further Reading:**
    Harrison (1980); *Inside Nature's Giants* | *PBS* www.pbs.org/show/inside-natures-giants/; Marshall Cavendish Corporation (2010) *Mammal Anatomy: An Illustrated Guide.*

a combination of them might be at stake in a given case of evolutionary change. We cannot simply assume that evolutionary adaptation is occurring at the simple level of the organism in a simple selection process in a given system. This fact has consequences for how we do evolutionary research and the research questions we ask, as we shall see in the upcoming sections.

3 Adaptationism and the Logic of Research Questions

3.1 Introduction

In "The Spandrels of San Marco and the Panglossian Paradigm," Stephen Jay Gould and Richard C. Lewontin (1979) drew attention to several risks of using an evolutionary research method that they called "adaptationism." They drew attention first to two weaknesses of adaptationism:

> [1]We would not object so strenuously to the adaptationist programme if its invocation, in any particular case, could lead in principle to its rejection for want of evidence. We might still view it as restrictive and object to its status as an argument of first choice. But if it could be dismissed after failing some explicit test, then alternatives would get their chance. Unfortunately, a common procedure among evolutionists does not allow such definable rejection ... [2] The criteria for acceptance of a story are so loose that many pass without proper confirmation. Often, evolutionists use consistency with natural selection as the sole criterion and consider their work done when they concoct a plausible story. But plausible stories can always be told. (Gould and Lewontin 1979, pp. 587–88)

Note here that in the first complaint, Gould and Lewontin are asking for a "stopping rule" of some kind that would signal the abandonment of the search for adaptive stories, or the time to investigate alternative accounts from the evolutionary canon.

The biggest danger they emphasized (issue [2] above) may be summed up briefly as the risk of what they called "just-so stories" serving in place of well-confirmed adaptive explanations. What happens in such cases is that the biologist's imagination serves her well for producing selective accounts for what are assumed to be adaptive traits, but that such accounts are not carefully examined for their empirical support, according to the standards described in Section 1.

Instead, the biologist's good storytelling ability substitutes for evidence in support of the adaptive account, and such "just-so stories" are accepted on the basis of their *plausibility* rather than firm evidential support for a variety of independent aspects of the model (see Section 1). In such cases, a variety of good evidence is neither sought nor provided; and the evolutionary basis of such traits is simply assumed and granted without firm evidential support – it is based on the attractiveness and plausibility of the conclusion or prediction of the evolutionary model.

A third problem clarified by Gould and Lewontin with adaptationism, as practiced in the manner they opposed, was the standard of assuming that selection was the only or the main evolutionary factor operating on a trait, thus neglecting the presence of the other evolutionary factors that we reviewed in Section 2, such as pleiotropy, phyletic inertia, and so on. These other factors, summed up as "evolutionary factors," are universally accepted as possible contributors to evolutionary change by every practicing evolutionary theorist but are often not considered as really "in the running" for actual explanations, complained Gould and Lewontin, and were given only "lip service." This meant, that even though these other evolutionary factors were recognized as technically possible, they were never acknowledged as *real possibilities* of evolutionary change, which they thought was a scientific error.

In this section, I present a framework for analysis that makes such "methodological adaptationism" and other ways to approach evolutionary research clearer.

Before we proceed, let us consider some definitional issues. Notably, there are other types of adaptationism, as highlighted by philosophers Peter Godfrey-Smith (2001) and Tim Lewens (2009). For example, there is *empirical adaptationism*, which emphasizes the predictive and explanatory benefits of the role of natural selection in evolutionary processes, and there is *explanatory adaptationism*, which claims that natural selection "has unique explanatory importance among evolutionary factors" (Godfrey-Smith 2001, p. 336). However, here I am concerned with *methodological adaptationism* understood as the view that "the best way for scientists to approach biological systems is to look for features of adaptation and good design" (Godfrey-Smith 2001, p. 337).

My focus is on subfields of evolutionary biology that rely on this approach in their day-to-day research, such as behavioral ecology, sociobiology, evolutionary psychology, and some branches of genetics, although it occasionally pops up elsewhere. In Section 4, I expand on the risks methodological adaptationism poses to scientific reasoning in evolutionary biology. Also, I emphasize that I am *not* attacking the notion of looking for adaptations in evolutionary studies: *I am not anti-adaptation*. The issues concern which framework of research questions is most appropriate and fruitful in any given case. But first, we must understand methodological adaptationism as a research program.

3.2 The Logic of Research Questions

As reviewed in the preceding sections, there are a variety of evolutionary factors that can influence the form and distribution of a given trait in a population or species. While natural selection at a variety of levels *may* be the most significant factor, we also have genetic linkage, phyletic history, developmental factors, embryological constraints, and social and niche coevolutionary factors, and, I would add, drift, mutation, chance, and migration, or combinations of any of these processes in a correlated or linked trait, among other causal and explanatory factors.

I call this basic approach using the full variety of evolutionary factors we've discussed the **evolutionary factors** framework of evolutionary theory; its fundamental research question is: **"What evolutionary factors account for the form and distribution of this trait?"** Sometimes, several of these factors operate simultaneously on a given trait, but often only one or two are the major factors causing its form and distribution at a given time.

When we investigate the evolutionary account of a given trait, in the fields we're discussing (i.e., evolutionary psychology, behavioral ecology, and sociobiology), it is usual to prioritize the functional factors, natural selection and sexual selection, as the most significant ones in evolutionary research. We might start with the question:

"Does this Trait Have a Function?"[20]

If the trait, after investigation, does not appear to have a correlation with fitness,[21] or does not appear to have evidence of design – hence, does not appear to have a current or past function – we pursue other possible evolutionary explanations, such as whether it might be due to genetic linkage with another trait or a by-product of selection (i.e., indirect selection). Alternatively, it may be present due to developmental or embryological constraints, or phyletic inertia, and so on. Pursuit of such explanations consists of testing them against available evidence and searching for new evidence specific to those factors.

However, the approach dominant among leading animal behaviorists, behavioral ecologists, many human evolutionists, and nearly all evolutionary psychologists (who ask evolutionary questions to analyze psychological traits), is the one described above (section 3.1), called **methodological adaptationism.** Under this approach, the leading research question is: **"What is the function of this trait?"** And the research consists of a search for supportive evidence for *adaptive* hypotheses that can explain the trait's presence in the population.

---

[20] See Rose and Lauder (1996) and Martins (2000) for examples and methodological details. Note that leading with this specific research question does not describe research in evolutionary developmental biology, which starts with nonfunctional questions in evolutionary developmental biology and leads with questions concerning form and structure.

[21] Symons gives reasons against using correlation with fitness for detecting adaptations and prefers evidence of design (1990).

Issues about method in behavioral ecology and biology revolve around evolutionary adaptations, one of evolution's biggest successes.[22] I take it as given that our living world is filled with such adaptations.

Let us examine a defense of an "adaptationist research program" by one of modern evolutionary theory's founders, Ernst Mayr, written in reaction to Gould and Lewontin's "Spandrels" paper. Mayr sets up the problem so that selection is the only positive answer to the evolutionary question. He writes:

> The adaptationist question, *"What is the function of a given structure or organ?"* has been for centuries the basis for every advance in physiology. If it had not been for the adaptationist program, we probably would still not yet know the functions of thymus, spleen, pituitary, and pineal. (1983 p. 328; emphasis added)

Mayr offers the discovery of the lateral line in fish – the complex dark line running along the sides of nearly all fish – as an example of the power of the function question; if we had just let the line remain mysterious, we would never have understood its function as an additional sense-organ in the life of fish.[23] Moreover, Mayr defends the adaptationist program as *harmless*, when applied correctly:

> The question whether or not the adaptationist program ought to be abandoned because of presumptive faults can now be answered. It would seem obvious that little is wrong with the adaptationist program as such, contrary to what is claimed by Gould and Lewontin, but that it should not be applied in an exclusively atomistic manner. (1983, p. 332)

Mayr's approach to adaptationism is widely adopted among evolutionary psychologists, behavioral biologists, and evolutionary ecologists alike.

Thus, *identifying the function* of a trait is the primary aim of the adaptationist program, which also aims to identify the formative selective pressures. Note that identifying a trait's function is not necessarily to identify its selection pressure: commonly, for example, we can have multilevel selection, and there are multiple processes responsible for the trait's form and function. The trait itself does not tell us how to describe its selection pressure, although the investigating biologist may play favorites about which process to privilege in their explanations.[24]

Note also that a trait having utility now is not the same as having a "function" in the selective sense. For example, a trait could contribute to

---

[22] See Lewens (2009) for nine types of adaptationism. Amundson (2001) and Sansom (2003) have also emphasized the multiple nature of adaptationist questions and answers, but not in the way I do here.

[23] See Williams 1966, pp. 10–11.   [24] For examples, see Wade 2016.

fitness in the current population without having been formed by selection to do so. Gould and Elisabeth Vrba (1982) dubbed such traits "exaptations"; these have current utility, or "effects," but not *functions* in the selective or engineering sense. Evolutionary psychologists and behavioral biologists don't often use this former category of evolutionary outcome, even when appropriate (see Section 4.4).

Alternately, a trait could have had a function in the past, and a correlation with fitness then, and be a "past adaptation," with no evolutionary function now, or even be an evolutionary "mismatch" now. Again, human and behavioral evolutionists rarely assign traits to past adaptation, and usually claim current fitness benefits.

While the adaptationist approach may look biased on its face, since it starts with the assumption that the trait is an adaptation rather than one of the other possible features, this assumption is supposed to be only temporary. If it turns out that the trait does not appear to have a function, then the biologist is supposed to move on to other possibilities. This supposedly benign program has been advocated by many biologists since Mayr.

Thus, the methodological adaptationist approach is seen as the "most helpful way to proceed": look for selective explanation in every case, which upon failure of the selective explanations might lead you to nonselective explanations, which could then be pursued if that is where evidence led (Godfrey-Smith 2001, p. 342). Adaptations are most often indicated by their "specificity, proficiency, precision, efficiency, economy, complexity of design, reliable production, costliness, etc." (Andrews et al. 2002a, p. 503 [from Williams 1966]). Optimality models, discussed in Section 1, can be used to investigate these features.

But it is still an open question whether the method outlined here *in practice* allows nonadaptive explanations to ever win the day. Remember that this issue of "lip service" was a key concern in Gould and Lewontin (1979). Do researchers *in fact* find themselves willing to embrace nonadaptive explanations when the evidence points away from adaptation? Other evolutionary factors are often explanatory, but are they appealed to when necessary under the methodological adaptationist approach?

I use what I call the Logic of Research Questions to help unpack and highlight the differences between the relevant theoretical approaches. The logic of the research questions we ask constrains what classes of answers we can give that are responsive and relevant.

Crucially, research questions carry with them an appropriate, unique class of possible and responsive answers, distinct from other contrasting classes of

> ### Box 3.1 "What is the function of this trait?"
>
> There are many possible and responsive answers to such a research question, all of which take the common form:
>
> *Possible and Responsive Answers*
>
> A: The function of this trait is F.
> A: The function of this trait is G.
> Etc.

answers. We need to think very hard about the research questions we ask and the answers they allow, because the questions can lead us to miss what's really going on, and therefore to scientific failure. While I apply this logic to the adaptationist methodology, it can be applied to any scientific investigation; the logic of research questions is applicable to any scientific field that experiences controversy about methods and inference.

The *methodological adaptationist* asks, echoing Mayr,

Someone following the *evolutionary factors* framework asks, on the other hand, quite generally:

We can now see a clear logical contrast between the two distinct frameworks and their corresponding sets of research questions and possible responsive answers.

Note that the first answer following the general question, "What evolutionary factors account for the form and distribution of this trait?", specifically, "Does this trait have a function?", is an adaptation answer, which was done to suggest that, pragmatically, in the fields of evolution we're examining, adaptation also is explored, as a priority, in the evolutionary factors methodology. This does not necessarily follow from the evolutionary factors logic, but it is a priority for those in the named fields of study because of their research interests, which center around functions.

But under the evolutionary factors approach, the key question about adaptation is: **"Does this trait have a function?"** which is *logically very different* from the key question asked by the methodological adaptationist. Here, there is **no assumption** that the trait is an adaptation, in sharp contrast to leading research with the question: **"What is the function of this trait?"** (A reminder for the reader: we are using Williams' and Lewontin's notion of an engineering function, not an etiological function; see Section 1.)

Under both the methodological adaptationist and the evolutionary factors approach, there are standards of evidence for when a claim for that factor is supported. When a claim is made that a feature is an adaptation, certain standards of evidence must be met, such as evidence of fitness contributions or design features. The same goes for evolutionary by-products/bonuses, or phyletic features; evidence must be given, in accordance with the standards discussed in Section 1.

Some of the arguments over adaptationism concern these standards of evidence, and nearly all of the focus has been on them, but I claim that some of the root issues concern the initial research questions.

Which question shall we start with? Is the methodological adaptationist question really harmless? Or is there something more being imported into the analysis with that question? Why not use the evolutionary factors research questions all the time?

3.3 Dangers of Methodological Adaptationism

3.3.1 The "Onerous Burden of Proof" and Its Disappearance

The methodological adaptationists routinely assume that some trait under consideration is an adaptation. Indeed, the research methods of adaptationism have proven very fruitful.

The burden of proof has always been on the adaptationists to demonstrate that a trait has a function and is an adaptation of one sort or another. G. C. Williams is usually quoted concerning the strong burden of proof required for an adaptive explanation:

> Demonstrating adaptation, Williams argued, carries **an onerous burden of proof**. Moreover, "This biological principle [adaptation] should be used only as a last resort. It should not be used when less onerous principles ... are sufficient for a complete explanation" ([Williams 1966] p. 11). Williams did suggest qualities of trait design that could help build a case for adaptation (e.g., precision, efficiency, economy) and claimed that formulation of "sets of objective criteria [of special design]" is a matter of "great importance ([1966] p. 9). (Andrews et al. 2002a, p. 496; emphasis added).[25]

But Williams himself applied only an informal probability standard for an adaptation: "whether a presumed function is served with sufficient precision, economy, efficiency, etc., *to rule out pure chance* ... as an adequate explanation (p. 10)."[26] So Williams himself allowed the "onerous burden" of showing adaptation to be satisfied, in practice, by something much weaker than the "objective criteria" he claimed were importantly needed.

Leading behavioral ecologists Hudson Kern Reeve and Paul Sherman (1993) also assume adaptation under a wide variety of circumstances, rejecting a widespread definition of adaptation, articulated by the leading philosopher of biology Elliott Sober, as being too weak. Sober requires, for an adaptation:

---

[25] There is an open question regarding how to read Williams' strictness on this topic (Lloyd 2013).
[26] Andrews et al. 2002a, p. 496; emphasis added.

"A is an adaptation for task T in population P if and only if A became prevalent in P because there was selection for A, where the selective advantage of A was due to the fact that A helped perform task T."[27]

Reeve and Sherman argue there is a big problem with history-laden definitions of adaptation like Sober's:

> Endler's (1986) survey reveals that ... knowledge [of selective history] is available for very few phenotypic attributes. This might mean that the majority of traits should be considered nonadaptations. *Alternatively, it might suggest the need for a new kind of definition.* (Reeve and Sherman 1993, p. 8; emphasis added)

Reeve and Sherman recommend counting more things as adaptations at the current time, using current utility: "An adaptation is a phenotypic variant that results in the highest fitness among a specified set of variants in a given environment" (1993, p. 1). Reeve and Sherman use current fitness to infer evolutionary history, that is, promoting to "infer evolutionary causation based on current utility" (1993, p. 2). They claim their approach "decouples adaptations from the evolutionary mechanisms that generate them" (1993, p. 1). This approach to adaptation is akin to what philosopher Bertrand Russell called the "method of 'postulating'," which he said "had all the advantages of theft over honest toil" (Russell 1919, p. 71). Thus, an additional problem with methodological adaptationism in practice is that it is prone to shirking its own "onerous burden of proof," that is, the standards of evidence that are set out in Section 1. It's easier to change the definition than it is to come up with the evidence of adaptation, so that is what some adaptationists do!

3.3.2 Mistaking Alternatives as Mutually Exclusive Rather Than Complementary Accounts

The logic of research questions under the evolutionary factors framework is a bit different from the logic described by many animal behaviorists, evolutionary psychologists, and other behavioral biology adaptationists, when they practice their craft of explaining the evolution of interesting organismal traits.

The evolutionary factors researcher sees the alternative evolutionary factors as potentially complementary, rather than seeing them as *mutually exclusive* to a selective approach. Thus, a given trait can be explained primarily through a selective force, but also through a genetic or developmental constraint on that selection, which narrows the range of selective results. In fact, combining

---

[27] Sober in Reeve and Sherman 1993, p. 7.

factors is very common among evolutionary biologists, as in Wright's combination drift and selection models (1931), and in the hierarchical selection models of Wade (1978, 2016) or Odling-Smee and colleagues (2003).

Under the evolutionary factors approach, rather than assuming that a trait is an adaptation, we can start our examination of any trait by asking: "Does this trait have a function?"

Also, the evolutionary factors approach acknowledges the division of scientific labor by saying it can be a good idea for some to start by asking about the function of a trait. Similarly, it would be more useful for a systematist to start by asking whether a trait is ancestral or "derived," and more useful for a developmental biologist to ask how the trait is developed in the organism.[28] The question we're facing in this chapter is: Is it useful for *anyone* to be a methodological adaptationist rather than following an evolutionary factors approach?

For example, contrast the evolutionary factors approach just described with this dualist methodological recommendation from Andrews et al.:

> Because hypotheses about constraint, exaptation, and spandrel, and hypotheses about adaptation are often **mutually exclusive** to each other, we have argued that *confidence* in these alternatives increases only when plausible adaptationist hypotheses have been considered, subjected to special design scrutiny, and *systematically rejected*. (Andrews et al. 2002b, p. 535; emphasis added in bold; original emphasis in italics)

Andrews and colleagues are making a point here about how to increase confidence in a hypothesis: if A and B are "mutually exclusive" exhaustive components, then increasing your confidence in component A commits you to decreasing your confidence in component B, and vice versa.

But the point is both logically and biologically mistaken, since the evolutionary components in question are not truly logically mutually exclusive: They can both be true at the same time. Although being an adaptation and being an exaptation are not mutually compatible, selection can also occur on a spandrel, making that combination also compatible.[29] Thus, Andrews et al.'s characterizing the factors as "mutually exclusive" is rarely correct in biology. To assert incompatibility is to misunderstand the selective and various other evolutionary factors. They are not only logically compatible, they are biologically compatible.

Mayr (1983) made the same mistake: "Only after all attempts to [find an adaptive explanation] have failed, is [the evolutionary biologist] justified in designating the unexplained residue tentatively as a product of chance," i.e., "the incidental by-product of stochastic processes" (Mayr 1983, p. 326; see

---

[28] Cf. Beatty 1987.   [29] See Table 1.1 for definitions.

> Box 3.2 "WHAT EVOLUTIONARY FACTORS ACCOUNT FOR THE FORM AND DISTRIBUTION
> OF THIS TRAIT," OR, E.G., "DOES THIS TRAIT HAVE A FUNCTION?"
>
> This question has a series of distinct possible and responsive answers (that
> might be considered in any order, except that the adaptive answers usually
> go first in practice for the fields under consideration):
>
> *Possible and Responsive Answers*
>
> A: This trait occurs in the population because it has the function F, an adaptation.
> A: This trait occurs in the population, it is under hierarchical selection,
> and it has the function G, an adaptation.
> A: This trait occurs in the population through mutation and genetic drift.
> A: This trait occurs in this population because of mutation and migration of genes
> from another population.
> A: This trait occurs widely in this population because of an evolutionary develop-
> mental novelty.
> A: This trait occurs widely in this population because it is genetically linked to
> a trait that is highly adaptive (genetic linkage or correlation).
> A: This trait has its current form largely because of an ancestral pattern (phyletic
> inertia).
> A: This trait has its current form and distribution because of pleiotropy with a trait
> that was under natural selection.
> A: This trait has its current form and distribution because it is a by-product or
> bonus of a trait that is strongly selected in the opposite sex in this species.
> A: This trait has its current form and distribution because of some combination of
> the above factors.
> Etc.

Millstein 2008). Similarly, Buss et al. (1998) narrow the alternatives to adaptation down to chance and "incidental byproducts," omitting all the other alternatives (see Box 3.2). This narrowing creates the appearance of having two
mutually exclusive disjuncts, but in fact, accumulating evidence for one does
not disconfirm all other evolutionary factors.

Modern evolutionary theory says there are evolutionary models available that
use distinct evolutionary causes; any of these may answer the question, "What
evolutionary factors account for the form and distribution of this trait?" Under
this analysis, the causes are not mutually exclusive; they can often be combined
and serve as complementary causes of evolutionary change.

The division of labor response says methodological adaptationists are merely
researchers who look for functional explanations for traits, and if they cannot
find one for a trait, move on to look for another functional solution for another
trait. If that were all they were doing, they would be quite harmless. But that is
not an accurate description of methodological adaptationism in practice. As we

will also see in Section 4, methodological adaptationism leads to bad logic, bad reasoning about evidence, and inferior biology. It is destructive of good science and good evolutionary biology; it is *not* just a matter of overemphasis on adaptation in the biological community.[30]

But there are even more serious problems that have arisen from methodological adaptationism. In practice, methodological adaptationists sometimes cannot compare the *weight of evidence* for various hypotheses, one against the other. Even when consideration of evolutionary hypotheses involving the other evolutionary factors really does happen, what counts as *evidence* supporting those hypotheses fails to come into view, as we shall see.[31]

4 Case Studies Showing the Inadequacy of Methodological Adaptationism

Introduction

We do not usually think that the logic of our scientific methods leads to closed-mindedness and the inability to see alternatives or evaluate evidence, but that is exactly what sometimes happens in evolutionary biology of behavioral and bodily traits with methodological adaptationism, despite its benign reputation.

In this section, we will look at some case studies in which adaptationist research programs compete with research programs that use other evolutionary explanations and research questions, focusing on my major case study on the evolution of the human female orgasm. It is claimed, by both biologists and philosophers who defend a methodological adaptationist approach, that adaptationist explanations will be abandoned once the evidence makes clear that such an adaptation approach will not work. I will examine why the research question "What is the function of this trait?" never seems to fail and never really rolls over into one of the alternative evolutionary explanations. I analyze why that is so. It is due to the logic of the research question and its possible and responsive answers.

4.1 The Case of Human Female Orgasm

4.1.1 Bonus/By-product Account of Female Orgasm

I have spent over 35 years researching and analyzing the evolutionary explanations for human female orgasm. In my 2005 book, I analyzed 22 distinct

---

[30] A similar point was made by Wagner in his discussion of shifting pluralism, which is "the idea that there are multiple causes and mechanisms involved in every evolutionary process" (2000).

[31] **Suggested Further Reading:** Amundson (1994, 1998, and 2005); Carroll (2005); Cosmides and Tooby (1994); Geary and Flinn (2001); Griesemer (2015); Griffiths (1996); Newman, S. A. (1988); Newman and Bhat (2008 and 2011); Pinker (1999); Raff (1996) Schmitt and Pilcher (2004); Symons (1990); Thornhill (1990); Wake (1991 and 2009).

theories – all that I could locate at that time – that offered evolutionary explanations for the existence and prevalence of female orgasm; 21 of them were adaptation accounts. I found that none of these 21 accounts were supported by the available empirical evidence that we had about female orgasm at that time.

Eleven of these adaptation accounts were based on a pair-bond view of female orgasm, while several other adaptive accounts were oriented around a female choice sexual selection view that depended on sperm competition and its concomitants (see Section 4.1.4). All of these adaptation accounts were attempting to account for the persistence of the trait in the human population, not its evolutionary origins per se, which might lie in the deep mammalian past, as postulated by Pavlicev and Wagner (2016). The alternative, nondirectly adaptive account for persistence is called the "by-product" or indirect selection account, and is based on strong stabilizing selection in males but not females, as explained below.

The evidential situation has only gotten worse for the 21 adaptive explanations for the prevalence of female orgasm since my book was published, because in 2013 a study of over 8,000 women was published in which the fitness difference that orgasm made for women was studied: The genetic component of fitness difference that the existence and frequency of orgasm made for women was a flat zero (Zietsch and Santtila 2013). In other words, having orgasm has no effect on genetic fitness, when measured as number of offspring.[32] If things in the past were like they are today, orgasm cannot have been an adaptation, since it made no difference to reproductive success, a necessary component for an adaptation. Is there any reason to think that things were different in relation to female orgasm in the past? Were pairs in the past more inclined to produce orgasm for the woman upon intercourse, as many have claimed to me, noting that modern women are "professional and uptight"? That argument would have to be substantiated with evidence, which does not look promising according to the world's authority on primate sexuality, Alan Dixson (2009, 2012).

I concluded in 2005 that the lone available indirectly selective account (which relies on substantial stabilizing selection in males), the by-product/bonus account, had the most evidential support, a position I still hold. The account was first proposed by anthropologist Donald Symons in 1979. An analysis of Symons' research shows that he is not a methodological adaptationist in the Mayrian sense, but rather an "adaptationist" following the path of the evolutionary factors framework question, "Does this trait have a function?" (See Symons 1990.)

Symons' account of female orgasm is based on the shared development of the embryo, as mentioned in the previous chapter. Males and females share the same

---

[32] Ideally, the fitness of female orgasm would be measured by both the quality and number of offspring, but this was not feasible for the Zietsche and Santtila study. Thus, what we have right now is a partial picture, but we have no reason to think it misleading.

embryological form at the beginnings of life – they start off with the same basic body plan, and only if the male embryos receive a jolt of hormones during the eighth week of pregnancy do any sexually distinguishing characteristics appear. In males, orgasms are understood to be stabilizing-selected adaptations, likely as rewards for participation in sexual intercourse, which raises reproductive success. In fact, while this is always assumed, it is supported only indirectly with evolutionary evidence of the kind we expect from adaptive accounts (Zuk 2006).

The understanding under the bonus/by-product or indirect-selective account of female orgasm is that the females get the orgasm "for free," that is, simply as a consequence of sharing the embryological form with the male, who is under stabilizing selection for orgasm. The tissues involved in orgasm for males and females are homologues, shared between males and females, including nerve tissues, erectile tissues, and muscle fibers. This whole embryological pattern, not just the clitoris (and potentially involving the five afferent [incoming] sensory nerve pathways I mention in my 2005 book), is involved in producing orgasm in females, and is produced in them through their embryological connection to the same tissues in males. The idea behind the theory is that the tissues involved in orgasm for males and females are very similar, including nerve tissues, erectile tissues, and muscle fibers. Thus, females get the functioning orgasmic tissues through this embryological connection and are capable of having orgasms under the right conditions of rhythmic stimulation.[33] Note that this may be seen as both an account of evolutionary origins of the trait, as well as a selective account of the trait's maintenance in the population.

The bonus, indirect-selective account may also be complementary to the Ovulatory Initiation Hypothesis of Pavlicev and Wagner, who hypothesize a deep origin of the female orgasm in the induced ovulation mechanisms, finding homologies in some of the hormonal releases of both features (Pavlicev and Wagner 2016, Pavlicev et al. 2019). I find it difficult to understand what the origins of the male orgasm are supposed to be under this account, but again, it may well be complementary to the by-product account, and they both have the same consequences for female orgasm today: Female orgasm is not adaptive, and it does not have a modern evolutionary function.

In any case, the indirect-selective sort of explanation for female orgasm is both 'developmental' and adaptive, as we reviewed in Section 2; female orgasms are seen as evolutionary "by-products" of adaptive evolution in the males of the species, just as male nipples are by-products or indirect-selective results of adaptive evolution in the mammalian females. When I first published my book, this more

---

[33] Komisaruk and colleagues have more recently shown that the human vagina, cervix, and clitoris are innervated by different afferent (incoming) pathways (2011). More research is necessary.

technical use of "by-product" language provoked such a backlash from feminists who accused me of belittling or dismissing the importance of female orgasm (despite my having researched and written a book for 22 years about it) that I have since changed the name to the "bonus/by-product" or more technically correct "indirect-selective" account, or sometimes, the "fantastic bonus" account.

Let us consider the variety of evidence relevant to this case. Note that we are considering supporting evidence *for* the by-product/bonus account. One common misperception is that supporting evidence *for* the by-product view must simultaneously be evidence *against* an adaptive view of female orgasm; but this is not this case and rests on seeing the by-product view as a null hypothesis, which it is not. We will discuss these matters later on in the section.

At the time that my book was published, we didn't have any fitness results from women's orgasm, as we do now, showing the lack of connection between fitness and orgasm. We did, however, know the distribution of women's orgasm rates with intercourse. Since vaginal intercourse is taken to be the behavior most highly correlated with reproductive success and fertility, orgasm rates with intercourse were taken to be significant. Figure 4.1 shows the distribution of orgasm rates with intercourse over a population of women; you'll notice that the curve is relatively flat, compared to the curves in the selective models shown in Figure 1.3.

This comparably 'flat curve' is represented by an *x*-axis of overall orgasmic performance, while the *y*-axis represents frequency in the population.[34] It shows that the orgasm rate with intercourse never rises above 20 percent for any segment of the female population, in contrast to the selection curves that you see in Figure 1.3 (in Section 1), where the curves feature large percentages of the population – along the lines of 70–90 percent – on the desired value of the trait. Only about 13 percent of women always have orgasm from intercourse (and about half of that is from "assisted" orgasm, that is, orgasm assisted by hand or vibrator stimulation during intercourse), and roughly a third of women rarely or never have orgasm with intercourse, while 5–10 percent of women never have an orgasm at all from any means. The rest fall between – sometimes do, sometimes don't have orgasm with intercourse. But, again, as we noted in Section 1 and saw illustrated in Figure 1.3, simple directional or balancing selection, which is supposed in nearly all of the selection stories of female orgasm, usually produces *substantial peaks* in the distribution curves of a trait, peaks encompassing 70–90+ percent of the population. (There is an exception to this expectation; see Section 4.1.4). The fact that women vary so widely in their orgasmic performance thus provides very suggestive evidence that nearly

---

[34] This curve was drawn from statistics from Dawood and colleagues 2005, but it conforms well with my analysis of typical numbers from the 35 studies that I reviewed in my 2005 book, which covered 66 years of research involving over 141,200 women.

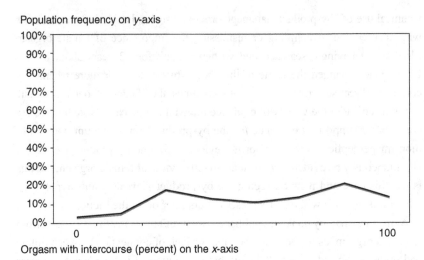

Population frequency on *y*-axis

Orgasm with intercourse (percent) on the *x*-axis

Figure 4.1 Orgasm rates with intercourse taken from Dawood and colleagues 2005

all of the selection explanations offered so far for female orgasm – no matter what their details are – are likely unsupported or disconfirmed. The most recent selective accounts published since my book (for example, Puts and colleagues, 2012a,b) do not address this issue of the lack of a selective peak.

Consider a further variety of evidence for the by-product/bonus account. Strikingly, females masturbate through direct stimulation of the clitoris – the homologous organ to the penis, from the same sources in the embryo – and not through simulating intercourse, just as we would expect on the by-product account. Similar stimulation to homologous organs yields orgasm for both sexes (Kinsey and colleagues 1953). Note the contrast with what we would expect about masturbation if orgasm were selected to go along with intercourse, as it does on *all* of the adaptive theories: Women would imitate reproductive sex when they masturbated.

In fact, only 1.5 percent of women do this, according to Hite (1976) and confirmed by Kinsey et al. (1953). Thus, the fact that women stimulate the homologous organ to the penis when they masturbate is predicted by the by-product/bonus account. The origins and maintenance of female orgasm lie with the male-homologous orgasmic structures and reflex (i.e., the clitoris and surrounding tissues), and not in relation to intercourse. The fact that this prediction of the by-product/indirect-selective account is confirmed is evidence in favor of the account.

These facts about masturbation are precisely why Freud thought it necessary to move the locus of excitement from the clitoris to the vagina for "mature" women, because it didn't make evolutionary or adaptive "sense." Freud believed that all women ought to have orgasm from vaginal intercourse. This belief was *explicitly derived* from his conviction that female orgasm must be an evolutionary *adaptation* to promote reproduction, like many present-day explanations of the trait.

Freudianism said: Female erotic pleasure is initially centered on the clitoris in infancy and childhood, but with the process of maturation into womanhood, the core of female eroticism must 'migrate' to the vagina as the healthy woman's 'primary' erotic center (Freud 1905).

As Freud wrote:

> In her childhood, moreover, a girl's clitoris takes on the role of a penis entirely: it is characterized by special excitability and is the area in which auto-erotic satisfaction is obtained. The process of a girl's becoming a woman depends very much on the clitoris passing on this sensitivity to the vaginal orifice in good time and completely. In cases of what is known as sexual anaesthesia in women the clitoris has obstinately retained its sensitivity. (Freud [1923] 1963, p. 318).

Women who failed to enact this migration – that is, the 90–94 percent of women who "obstinately" remained dependent on the clitoris for orgasm, or who failed to reliably have orgasm from intercourse without additional clitoral stimulation – were 'infantile,' 'immature,' 'neurotic,' 'frigid,' or 'dysfunctional.' Each of the labels is an assault on women's well-being itself. It's very significant that this labeling was not restricted to some medicalized or head-shrunken subset of the population – it was the coin of the realm, in women's magazines and advice columns, in doctors' advice, and in marriage manuals.

But rather than follow the other evolutionary explanations in enforcing this abuse, the bonus/by-product indirect-selective account actually explains the low rate of reliable female orgasms with intercourse.

It is also supported by new anatomical studies linking the structure of the genitals with orgasm rate with vaginal intercourse. Behavioral endocrinologist Kim Wallen and I discovered that women with a longer distance between their clitoris and urinary meatus (urinary opening, as a more stable substitute for a measure of vaginal opening, the CUMD, "Clitoral-Urinary-Meatus-Distance") have many fewer orgasms with intercourse than those women with a shorter distance.[35] Note again that the occurrence of orgasm is not correlated with fitness measures, so these different distances cannot be interpreted functionally, under the present information. The correlation we discovered seems to be an accident of development, not a product of genes, which is confirmed

---

[35] Wallen and Lloyd 2011; independently confirmed through MRI by Oakley et al. 2014; Vaccaro 2015.

independently by animal studies (Wallen and Lloyd 2011; see discussion and Figure 4.2 in Section 4.1.4).

Symons's general thesis is also supported by the nonhuman primate evidence. For example, female stumptail macaques have been shown to have orgasms by placing electrodes into their uteruses and genitals – which shows that they have the distinctive contractions and other bodily markers characteristic of orgasm.

There is a fascinating thing about these orgasms, though. Many of them occur when one female climbs onto the back of another female and rubs her clitoris against the other one's back, thus providing the direct, rhythmic stimulation desired for orgasm. A few of these same females also have orgasms while copulating with males, but these rates tend to be quite low (Lloyd 2005a). In addition, none of the observations about homosexual orgasms were observed to occur during the monkeys' fertile periods, thus denying any account based on hormones.

All of this is bad news for evolutionary theorists attempting to tell adaptive stories of female orgasm that tie it to heterosexual copulation or fertility, with which this evidence is incompatible. For example, the adaptive hormone-based account of female sexuality says that orgasm will occur during the estrus period of the animal (i.e., when it is fertile), but this is not what is found here, where the orgasms all occurred outside the fertile period, and cannot therefore be related to fertility. These orgasms, however, that are unrelated to heterosexual copulation and to fertility are completely compatible with Symons' by-product/bonus account, as they link the clitoral stimulation to penile stimulation of the male. This behavior is not compatible with any of the proffered adaptive accounts of female sexuality (Lloyd 1993).

4.1.2 Adaptationist Responses to the Bonus/By-product Account of Female Orgasm

In a 1987 discussion in *Natural History* of my early work presented by Stephen Jay Gould, on the evolution of female orgasm, Gould discussed a variety of empirical evidence in favor of Symons's by-product/bonus view. Adaptationist Donald Dewsbury, a distinguished psychologist studying animal reproductive behavior, claimed in response to Gould's discussion

> But Gould (1987a) goes too far in asserting that "female orgasm is not an adaptation at all" (p. 17). We need to study the consequences of [female] orgasm for differential reproductive success and then determine whether a plausible case can be made for drawing the loop from present consequences to the past history of natural selection. These need to be studied, *not asserted or denied a priori.* (Dewsbury 1992, p. 103; my emphasis)

But Gould is actually representing Symons' views in his quote, and Symons' views are based on empirical data presented in his chapter; they are not asserted a priori. Gould's full quote says: "In all the recent Darwinian literature, I believe that Donald Symons is the only scientist who presented what I consider the proper answer – that female orgasm is not an adaptation at all. (See his book, *The Evolution of Human Sexuality* (1979))." (Gould 1987a, p. 17).

Gould also says:

> Elisabeth Lloyd, a philosopher of science at the University of California at San Diego, has just completed a critical study of explanations recently proposed by evolutionary biologists for the origins and significance of female orgasm. Nearly all these proposals follow the lamentable tradition of speculative storytelling in the a priori adaptationist mode. (1987a, p. 17)

But Dewsbury's perception was, clearly, that no good evidence had entered into the debate, despite Symons's entire book chapter detailing empirical evidence supporting his theory, and Gould's appeals to the empirical support that I had amassed, involving 14 studies at that time. But all of that empirical evidence (discussed in Section 4.1.3, "'Null' Hypotheses") was *invisible to these researchers*. Apparently adaptive hypotheses could be favored or disfavored by the evidence – and they had not been favored in the female orgasm case so far – but a nonadaptive hypothesis like the by-product/bonus account apparently could only be "asserted or denied a priori."

Similarly, many years later, evolutionary psychologists Andrews and colleagues, in considering Gould's discussion of the female orgasm, claimed that Gould gave no positive evidence for the trait not being an adaptation, under either the contemporary fitness view of adaptation or under a historical functional account, and simply "proclaimed that the female orgasm is *not* an adaption but a byproduct."[36] Andrews et al. complain about Gould's methodology that:

> Gould's conclusion may be correct but his argument does not warrant it. As we point out later, demonstrating that the female clitoris and orgasm are byproducts requires the failure to find evidence for its special design and, hence, an adaptationist testing strategy. (Andrews 2002a, p. 499)

Note Andrews and colleagues' requirements of the evidence for something to be shown to be a by-product/bonus. They complain that Gould did not fulfill their evidential requirements, which did not concern positive evidence for the by-product account, but rather, failure at finding an adaptation.[37]

---

[36] See Box 4.5 to see the rankings of the strength of evidence and sources for these features of the evidence for the indirect-selective by-product account.

[37] This is a summary of supporting evidence, not a summary of evidence against an adaptive account (see Section 4.1.3).

Box 4.1 Variety of Evidence Supporting the By-product Account[38]

Developmental homology between penis and clitoris,[39]

Developmental homologies between male and female nerve fibers and pathways, and erectile tissues

Female masturbation techniques focused on the clitoris, not reproductive sex (intercourse)

Low rates of female orgasm with reproductive sex

Female orgasm in nonhuman primates (outside of intercourse and estrus)

Correlation of unselected female anatomical traits with orgasm with intercourse[40]

But actually, the repeated failure of adaptationist accounts does not have any bearing on the *positive evidence* available supporting the bonus/by-product account (reviewed in Box 4.1), although many adaptationists incorrectly believe that this failure of the adaptationist accounts is the sole evidence supporting that indirect selective explanation.[41]

As we'll see later on, because the logic of the methodological adaptationists' research question demands exclusively a function answer, no bonus/by-product answer can be considered a positive answer to their research question, and thus bears support in its favor.

But what about the other evolutionary factors that are allowed in all evolutionary textbooks? Reeve and Sherman (1993) allow that there are "mechanisms of persistence other than natural selection," and they list them as follows: "a relatively non-adaptive trait may persist because of several processes including prolonged lack of genetic variation, unbreakable genetic correlations with other traits, recurrent immigration, and genetic drift" (Reeve and Sherman 1993, p. 19). But it turns out that in practice they take these alternative causes, such as developmental constraints or genetic correlation, to either actually be serving the adaptive functions as well, or to not really be viable as *alternative causal explanations* to adaptive explanations. From this we can see why the list of answers to the methodological adaptationist research question does not actually include *any other answers* besides the function ones.

---

[38] See Section 4.1.4.   [39] Andrews et al. 2002a, p. 499 footnote 6: p. 504.
[40] Alcock 1987, 1998; Sherman 1989; see Lloyd 2005a for discussion.
[41] See Alcock 1998; Linquist 2006; Mayr 1983, p. 326.

4.1.3 "Null" Hypotheses

Ritual recitation of Gould and Lewontin's "Spandrels" paper in the adaptationist literature usually includes only the lesson that not *everything* is an adaptation. But this misses one of the primary points of their paper, which includes the problem of the neglect of developmental constraints, phyletic inertia, and laws of form as evolutionary causes. Attention to the logic of research questions illuminates this problem. How can these other factors ever appear on the methodological adaptationist's list of real, responsive answers?

I argue that methodological adaptationists are committed to this neglect by the logic of their initial orienting question. Once this first move of committing to the question "What is the function of this trait?" is carried out, which seems so innocent, methodological adaptationism is rationally going to lead to error in some cases. This is because starting our biological inquiry by asking the methodological adaptationists' function-question involves treating nonadaptive hypotheses as something like statistical nulls.

For example, David Barash says explicitly, in a discussion regarding the by-product/bonus theory of female orgasm, that the possibilities include "the 'null hypothesis' that it might not be a direct product of evolution after all" (2009, p. 133).[42]

In general usage in science or biology, a "null hypothesis" is usually a negative alternative to a positive correlational hypothesis, often used in Neyman-Pearson statistical analyses, one that binds together two variables or terms in a pattern of relations. The use of the "null" hypothesis by behavioral biologists tends to be much more informal, and not to signify necessarily the application of any formal statistical test at all. The positive hypothesis in this case would be one in which a genetic or phenotypic trait was positively correlated with fitness or some component of fitness, while the null hypothesis would be simply a noncorrelation with fitness, often indicating nonselection. An example will bring this out.

John Alcock is the author of the leading textbook on the evolution of animal behavior, but he writes about Gould, who is following my nonadaptive approach to female orgasm: "These are not the claims of someone who wishes to expand the horizon of evolutionary analyses but instead are designed to marginalize the adaptationist approach." Alcock is thus equating *evolutionary causal analyses* with adaptationist analyses exclusively (1998, p. 332). This neglects the fact that the by-product account of female

---

[42] By "evolution," Barash means "selection," in context. This mistake is discussed in the next section.

orgasm by Symons is, in fact, an evolutionary causal analysis in its own right.

In another paper, Sherman simply assumes what needs to be shown:

> Alcock (1987) had good reasons to question how a structure that plays such an intricate role [in facilitating sexual pleasure in women—here, he is talking about the clitoris, not the orgasm, inappropriately] and one *so obviously related to fitness* could possibly be reproductively neutral. (1989, p. 698, emphasis added)

The adaptationists continue to behave as if there's no supporting evidence for the by-product view, seeing the view as being essentially negative: They see the by-product view as claiming only that no adaptive account has been found, rather than as a positive causal story of indirect selection and its by-products in its own right. Hence their attitude that adopting the by-product view as a live alternative to an adaptationist account of the trait is tantamount to "giving up" on an evolutionary account altogether.[43]

This, however, is the wrong standard scientifically. It presents the scientific situation as all-or-nothing, the adaptation account or no scientific account at all. The by-product/bonus theory *is* a causal evolutionary account and has its own kind of evidence in its favor (see Beatty 1987 for a parallel point about genetic drift). Analysis of the logic of research questions helps make the source and nature of this confusion quite clear.

Let's review the questions asked by the methodological adaptationists and the researchers using the evolutionary factors framework and their samples of relevant well-formed answers here.

As can be seen by analysis of the logic of these research questions in Box 4.2, then, those using the methodological adaptationist approach cannot appreciate the accumulated positive evidence (see Box 4.1) for the by-product/bonus approach, because they take it as equivalent to a null hypothesis. The by-product account involves positive stabilizing selection on the males, and embryological evidence supporting the linking of the relevant trait with the females of the selected trait, among other evidence, as listed in Box 4.1. This evidence is in some sense *only visible on the evolutionary factors framework*, where the weight of evidence is the right approach to use in evaluating the by-product/bonus causal hypothesis and its alternatives.

---

[43] For example, see the language in Alcock 1987, 1998; Sherman 1989. "The adaptationist position is an invitation to scientific investigation," writes Alcock, and this is meant to contrast with the by-product account, which, it is implied, stifles inquiry prematurely, resting with an unsatisfactory, incomplete explanation (1987, p. 6; Lloyd 2005a p. 156 and Ch. 6).

BOX 4.2 ANALYSIS ACCORDING TO THE LOGIC OF RESEARCH QUESTIONS

Methodological Adaptationists: "What is the function of this trait?"
Possible and Responsive Answers

A: The function of this trait is F.
A: The function of this trait is G.
A: The function of this trait is H, or I, J . . . Z, AA. . . ?
Faulty, Nonresponsive Answer: This trait has no known function or correlation with fitness, and may not be a direct product of selection at all (i.e., adaptationists' "null" hypothesis, which Barash treats as equivalent to the bonus/by-product hypothesis)

Evolutionary Factors Approach: What evolutionary factors account for the form and distribution of this trait? Does this trait have a function?
Possible and Responsive Answers, often used in combination:

A: This trait occurs in the population because it has the function F, an adaptation.
A: This trait has its current form and distribution among one sex largely because it is a by-product or indirect product of stabilizing selection and genetic linkage on the opposite sex's trait. (The methodological adaptationists' mistaken "null hypothesis" actually is an independent causal hypothesis, which can have its own positive evidence.)
A: This trait occurs widely in this population because it is genetically linked to a trait that is highly adaptive.
A: This trait has its current form largely because of an ancestral developmental pattern that provides a contemporary developmental constraint.
A: This trait has its current form and distribution largely because of a phyletic pattern from an ancestor that is continued in this trait.
A: Combinations of above factors
A: etc.

Two points emerge from the discussion above.

First, characterizing the by-product/bonus alternative as a "null" hypothesis leads to the impossibility of positive evidence for what is, in truth, a causal hypothesis, which needs empirical support or refutation. Thus, the attribution of a "null" is mistaken.

Secondly, both the methodological adaptationist and the evolutionary factors theorist can ask about adaptive traits or functions, but the full meaning of the question will not be revealed until we can see what list of answers are *live options* and under full consideration. As we have seen, the methodological adaptationists may claim to have the same items on their lists as the evolutionary factors theorists, but when push comes to shove, they do not treat them as live options, on their own accounts.

We can zero in on the key problem: None of this evidence for the causal by-product/bonus hypothesis in Box 4.1 is recognized or weighed when considering this trait from the adaptationist methodological perspective. Thus, even though the methodological adaptationists present their adherence to their research program and its attendant question as perfectly harmless and, in fact, very good and productive science, we can see here an example of where it goes astray. Because they implicitly assume that no null hypothesis may have evidence in its support, they **cannot see the evidence** supporting the by-product/bonus account.

4.1.4 Critics of the By-product Account and Evidence

The leading critics of the by-product/indirect selection account continue to support adaptive accounts. Let us compare the status of the empirical evidence supporting the most favored current adaptive account (Puts et al. 2012a,b, Box 4.4) with the evidence supporting the bonus/by-product account, as given in Box 4.1. Recall that, according to 37 studies of 148,346 women using 27 metrics in the various studies, only about 20 percent of women reliably have orgasm with intercourse, including with hand-or-vibrator-assisted intercourse, while about 90 percent of the female population does have orgasm sometime during their lives (Lloyd 2005a). So, while orgasm is present in the vast majority of women, it does not routinely appear in the ordinary evolutionarily relevant context, that is, vaginal intercourse.

The by-product/bonus account gives a cogent explanation for this glaring fact. However, we should note that not every evolutionary explanation must produce frequent orgasm with intercourse, under our current theories. Consider a female choice account with sperm "upsuck," as shown in Box 4.3.

Puts and colleagues' female choice/upsuck account posits that females mate multiple times over a short period of time with different males. The basic idea is that the female will have orgasm preferentially with the higher-quality males. Orgasm is assumed to be accompanied by a mechanism of uterine upsuck that makes it more likely that the female will be fertilized by the higher-quality male. Thus, the orgasmic women are required to respond with orgasms only sometimes with intercourse – "yes" with high-quality males, and "no" with lower-quality males. The model says this will produce more and/or better offspring. In Box 4.3 there is the sketch of the female choice selection model and its level of evidence. But look at their data (Box 4.4).

> Box 4.3 Female Choice/Upsuck Model of Female Orgasm and Its Empirical Evidence
>
> Trait – [Mate with multiple males]: **Fair evidence**[44]
> Trait – [Female orgasm preferentially with high-quality males]: **Fair evidence**[45]
> Trait – [Uterine upsuck]: **Evidence against**[46]
> Hereditary basis – [Orgasm's heritability]: **Good evidence**[47]
> Connection to fitness – [More and better offspring with higher orgasm frequency]: **Evidence against**[48]
>
> Selection pressure – [Strong pressure on women to have offspring of high-quality fathers; female choice, sperm transport]: **Poor evidence**[49]

Alan Dixson, again, the world authority on comparative primate sexuality, denies that this sort of "cryptic female choice" selection occurs in human beings (2012, p. 630), in which females have a mechanism to "choose" or prefer the sperm of higher-quality males to fertilize their eggs. Dixson gives a series of reasons why this is not so, including the lack of specialization in the penis, the lack of specific structures in the sperm, the lack of evidence for uterine upsuck,[50] and so on, all of which, in other species of primates, accompany cryptic female choice and sperm competition as hypothesized in the upsuck theory, and its associated hypothesis of female choice.

Under the upsuck theory, female orgasm is accompanied by a sucking motion of the uterus. This "uterine or sperm upsuck" theory is an old piece of folk wisdom dating back to the Ancient Greeks, but it has never enjoyed any real evidential support, although people have been seeking scientific evidence to support it for over 70 years. There is actually evidence **against** it from a set of nicely designed experiments done by Masters and Johnson in the 1960s (1965; 1966)

Nevertheless, the upsuck hypothesis is very widely believed by laypeople, especially due to a deceptive narration of a video clip shown on TV,[51] and had achieved widespread acceptance among scientists since the 1990s, through the work of Robin Baker and Mark Bellis published in 1993, which claimed to provide empirical evidence supporting the phenomenon.

---

[44] Puts et al. 2012a,b.   [45] Puts et al. 2012a,b, giving them the benefit of the doubt.

[46] Levin 2011; Masters and Johnson 1965, 1966; Dixson 2009; 2012.

[47] Dawood et al. 2005; Dunn et al. 2005.   [48] Zietsche and Santtila 2013.

[49] Puts et al. 2012a,b. See Lloyd 2005a, Ch. 7; Levin 2011, 2015; Dixson 2012 for critiques.

[50] See Levin 2011 for further evidence against sperm upsuck.

[51] While the video only showed the contractions of the uterus upon orgasm dipping down into the vaginal cavity, where the sperm lay, the narrator added that "upsucking" of the sperm was also happening, although we cannot witness that.

Box 4.4 *Visualization of* Baker *and* Bellis 1993 *Data Set from Pairs in Study*

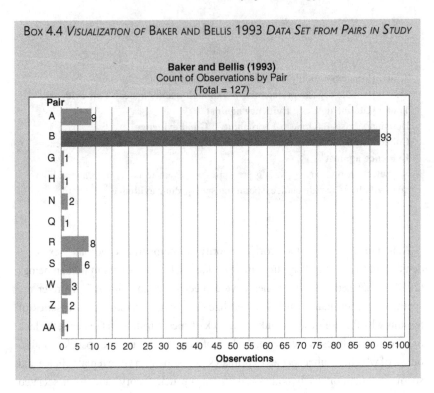

In one "supporting" data set, they have 1 out of 11 couples in the sample contributing 93 out of the 127 data points (nearly three-quarters of the data!). Four of the other 10 couples contributed one data point each, a combined total of 3 percent of the data, and so on. But extrapolating to the population at large based primarily on the results of a single subject badly violates standard statistical practice (see Lloyd 2005a, Ch. 7). In the end, the Baker and Bellis data are statistically worthless and no scientific conclusions can be drawn from them.

So why was this account of uterine upsuck adopted by top researchers in human evolution and taught in human evolution courses by evolutionary psychologists and sociobiologists? It cannot be because the paper was good science – because it clearly wasn't – as even my harshest critics now agree (e.g., Barash 2005). It seems to be because they just *liked* what the paper had to say, namely that female orgasm serves an evolutionary *function*. Here we see adaptationist bias in action, and several adaptationists, David Puts and Khytam Dawood and colleagues, are *still citing* this Baker and Bellis paper to underpin their female choice theories (e.g., Puts et al. 2012a, b) despite having given no real defense of the paper against the critique I gave in 2005, among other methodological critiques.

In Box 4.3 is the state of the empirical evidence supporting the various aspects of the female choice/upsuck selection model, accompanied by their sources. As you can see from Box 4.3, the model is not well supported by the empirical evidence.

But if we look just at the *prediction* of this particular sexual selection model, it can successfully be tweaked to fit the evidence of the flat curve showing orgasm rate with intercourse, as in Box 4.1. The poor state of evidential support for the hypothesis is only made clear by checking the independent evidence for the various aspects of the model, the central and crucial step in evaluating an adaptive hypothesis, as we saw in Section 1. Thus, the only adaptive model that is compatible with the sexology evidence – in that it fits the flat curve – does not have further supporting evidence for its aspects and assumptions. Thus, the female choice model is not comparable to the by-product account in its overall evidential support.

There is further evidence in favor of the by-product account. One of the most compelling things about the female choice theory is that it gives an answer to the question "Why do women respond to intercourse with orgasm sometimes yes, sometimes no?", which was mysterious under other adaptive theories.

Kim Wallen and I published independently confirmed analyses of data that bear directly on whether or not women have orgasm with intercourse, and when. As mentioned above, our anatomical, *non*-adaptive, explanation of why women have a variety of orgasm rates is based on what we call "CUMD" measurements, the measured distance between the clitoris and the urinary opening, meant to stand in as a proxy for the vaginal opening.

Figure 4.2 shows some of our data.

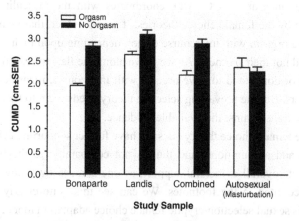

Figure 4.2 Wallen and Lloyd CUMD

The white bars represent women who reliably *do* have orgasm with intercourse, while the black bars represent women who routinely do *not* have orgasm with intercourse. On the *y*-axis is the CUMD.

You can see that the women who routinely have orgasm with intercourse have much shorter CUMD measures than the women who do not. The difference between the groups in orgasm rate is highly significant, greater than two standard deviations, in both data groups that we analyzed, Bonapart and Landis. This anatomical trait is apparently due to developmental exposure to androgens in the womb and is not genetically controlled.

Using an independent statistical test, we also found that this anatomical *distance* was predictive of whether a woman had orgasm with intercourse. You can also see that it was irrelevant to orgasm with masturbation, as we would expect.

In other words, we found that an anatomical trait predicted whether or not a woman would have orgasm with intercourse. This makes good sense, if you think about it. In sum, Wallen and I found that for the strong majority of women in our samples, anatomy seems to suggest destiny (2011). And clearly, if her anatomy so strongly influences whether or not she has an orgasm with intercourse with a male, that leaves insufficient room for the genetic quality of the male – specifically, his masculinity, attractiveness, or dominance, as presented in the female choice theories – to strongly influence the outcome of such intercourse, with orgasm accompanied by uterine upsuck.

This is because, according to the population genetic theorists, the selection pressure of this type of sexual selection scenario needs to be *quite strong* in order to produce any result in terms of evolution (Hosken 2008). Thus, the anatomical relation we discovered makes such a scenario highly unfeasible. In other words, there is no evidence that the women's variation in orgasmic response, as measured in CUMD, coordinates with the masculine features emphasized by the female choice theories. It seems, rather, that women either tend to have orgasm with intercourse or not, depending upon their anatomical features and not their partners. As we move along the flat curve, those women either tend or don't tend to have orgasm with intercourse, regardless of their male partners. But the prevailing selective theory needs them to be very sensitive to male traits against the available evidence.

Thus, the female choice theory does not have further supporting evidence for its aspects and assumptions, and thus is not comparable to the by-product account in its overall evidential support, as one can see by following up on the evidence cited in the footnotes. We can see this quite clearly when we contrast this sexual selection/cryptic female choice adaptation model (Box 4.3) with the one for the bonus/by-product/indirect selective account (Box 4.5).

BOX 4.5 EMPIRICAL EVIDENCE FOR THE BY-PRODUCT/BONUS HYPOTHESIS

Trait – [Developmental homologies between orgasmic tissues]: **Excellent evidence**

Trait – [Effectiveness of female masturbation in producing orgasm]: **Excellent evidence**[52]

Trait – [Low rates of female orgasm with intercourse]: **Excellent evidence**[53]

Trait – [Female orgasm in nonhuman primates]: **Good evidence**[54]

Hereditary basis – [Orgasm's heritability]: **Good evidence**[55]

Connection to fitness – [NO correlation with fitness]: **Good evidence**[56]

Note that in both cases, there is evidence for *no function* of orgasm in females, which is *nonresponsive* to the research question of the methodological adaptationists, "What is the function of this trait?" Instead, the by-product /bonus explanation is better seen as an answer to the evolutionary factors research question, "What evolutionary factors account for the form and distribution of this trait?"

This is the correct reading of the bonus/by-product theory of female orgasm, as a positive alternative causal hypothesis, not as a null hypothesis. It is an alternative to the previous, function answers to the methodological adaptationists' question, and it is an answer that is *not on their list of possible answers*, which only includes answers like: "The function of female orgasm is to preferentially mate with high-quality males," *or* "The function of female orgasm is to aid the pair bond," and so on.

To illustrate the dangerous and unscientific consequences of methodological adaptationism in this particular case: Several prominent adaptationists repeatedly complain that under the by-product/indirect selection hypothesis, female orgasm would fade away and deteriorate over evolutionary time and would disappear from the population. This mistaken inference has been advanced not only by leading scientists such as Alcock, Sherman, and Barash, but also by feminist primatologist and human evolutionist Sarah Blaffer Hrdy, and it is based on a misunderstanding of both how the by-product/bonus account works, and the evolutionary factors framework itself.[57] These misunderstandings are

---

[52] Kinsey et al. 1953; Hite 1976; Fisher 1973.

[53] Kinsey et al. 1953; See summary of 47 studies, 140,000+ women in Lloyd 2005a.

[54] See summary in Lloyd 2005a; Lemmon and Allen 1978; Chevalier-Skolnikoff 1974.

[55] Dawood et al. 2005; Dunn et al. 2005.

[56] Dawood et al. 2005; Dunn et al. 2005; Kinsey et al. 1953; Komisaruk et al. 2006; Lloyd 2005a; Zietsch and Santtila 2013.

[57] Alcock 1998; Sherman 1989; Smith (Hrdy) 2005.

likely a consequence of their adaptationist bias that a particular trait will only be sustained in a population if it itself is under sustained selective pressure. But under the bonus/by-product account, the basic muscle, nerve, and tissue pathways involved in female orgasm would be maintained in the female over the generations *in virtue of the fact that they are under ongoing strong stabilizing selection in the male*; male nipples are maintained in the same fashion. Nobody thinks that male nipples are disappearing! Thus, methodological adaptationist explanatory biases involving the necessity of selection have led to fundamental mistakes regarding the by-product/bonus hypothesis by these researchers and their followers.

In addition, when discussing alternatives to adaptations, methodological adaptationists have been prone to make further scientific errors concerning what the by-product/bonus account says and assumes. These biologists reason that if a trait is not *adaptive*, it cannot be part of an *evolutionary* account at all.

On John Alcock's analysis, the by-product/bonus hypothesis is a null result and offers only a "proximate" explanation of how women come to have orgasms.

For Alcock, who calls himself an "ardent adaptationist," the by-product explanation is seen as no evolutionary explanation at all. On Alcock's analysis, which is shared by Paul Sherman, a leading theoretician of animal behavior, the by-product hypothesis offers only a "proximate" explanation of how women come to have orgasms. In other words, it only explains how female babies grow up to have orgasms as adult women.

Alcock writes that

> proximate explanations of a biological characteristic do not make it impossible to ask whether the trait of interest contributed to individual reproductive success in the past or does so currently ... If we were to discover the female orgasm occurred with positive effects on female reproductive success, we would gain an *evolutionary* dimension to our understanding of this trait that is not covered by *any* proximate explanation. (emphasis mine, emphasis his, Alcock 1998, p. 330)

Thus, the by-product account is not seen as an evolutionary account at all – it is not an answer to *any* evolutionary question about female orgasm, with supporting evidence and theoretical standing in evolutionary theory.

Here we can also look to David Barash, the author of the most widely selling textbook on sociobiology for a couple of decades and a grandfather of the field of human evolution, who writes, with his wife, in a sympathetic discussion regarding the impetus behind those favoring the by-product/bonus theory, that it involves a scientifically legitimate desire to explore all possible explanations for any biological enigma of this sort, including the "null hypothesis" that it might not be a direct product of *evolution* after all. (2009, p. 133; my emphasis)

Note the equivalence of evolution with selection in this statement; the bonus/by-product explanation is mistakenly not considered *evolutionary*, just as we saw before with Alcock and Sherman. This is again the result of the logic of the research question, through methodological adaptationism.

For these authors, unless we are allowed to assume there is an adaptation, then we cannot tell whether we can explain it in an "evolutionary" way. The entire rest of evolutionary biology that we have been discussing in the evolutionary factors methodology is invisible, under this account; *it is disappeared*. The methodological adaptationists' methodology was supposed to be benign; it was not supposed to be a risky endeavor with radical theoretical commitments, although that is where it seems to have ended up.

4.1.5 Beyond the Comfortable Boundaries of Adaptation: The Logic of Research Questions

In this section, I have been emphasizing the initial patterns of inference and explanation, of exploration and investigation, rather than the final "evidentiary standards," that are often emphasized and discussed when considering adaptations. My focus here is much more on *the investigative standards*, and less on the evidential standards. My point is that if you use the methodological adaptationist research question, the evidentiary standards of alternatives like the by-product/bonus view, accurately portrayed, never come up, since they are buried under the assumptions regarding the null hypothesis and other myths. Gould and Lewontin's just-so story objection is about the standards of evidence, but I have identified the deeper danger earlier, which is in the logic of the research questions asked, particularly in the consequences of the methodological adaptationist research question and its possible and responsive answers all being "function" answers.

Many adaptationist researchers approaching the evolution of female orgasm start with the methodological adaptationist research question:

"What is the evolutionary function of female orgasm?"

Other evolutionists, though, may ask:

"What evolutionary factors account for the form and distribution of female orgasm?"

My claim is that the possible and responsive answers to these evolutionary questions are relevantly and significantly different. These are questions and answers with consequences.

For instance, an evolutionary account of female orgasm is used to underpin a notion of the normal function of female orgasm for the diagnostic and statistical manual of the psychiatric profession, the DSM V by which sexual disorders are diagnosed.[58] But in evolutionary science, there is no single accepted adaptive account of female orgasm, and thus no notion of its 'normal function,' if it has any at all.

The fact that the adaptationists see the by-product view only as a null hypothesis or as a nonanswer to the adaptive question leads directly to their mistaken characterizations and inferences involving the view. I am concerned with what it would take to get a community of adaptationists to move far enough beyond their research orientation and their driving adaptationist question to start approaching things from the broader, evolutionary factors point of view.

What would it take for these researchers to consider seriously the alternatives of the evolutionary factors approach, which are acknowledged to be real contenders for valid evolutionary explanation?

This case of the female orgasm reveals that there is something peculiar and implausible about methodological adaptationism as it's usually advanced, namely, that the researcher is envisioned as, at some time in the middle of their research program, abandoning their research commitments and explanatory practices, in the face of some facts or others, and *then going along some completely different explanatory pathway.* This is a lot to ask a researcher to do; they need to do a sort of mental gymnastics, and few – in fact, none – of the methodological adaptationist researchers involved with the evolution of female orgasm have been able or willing to do that.

In sum, this case highlights issues with the evidence needed for adaptive explanations that are pertinent to the study of development and by-products, and how confusing to biologists it sometimes is to move beyond the comfortable boundaries of assuming that a trait is an adaptation.

Let us consider some other cases in which developmental biology and developmental genetics play a significant role in evolutionary explanation and examine how adaptation explanations might work in counterpoint to evolutionary factors ones.

4.2 Development and the Case of the Salamander Toes

Recall from Section 2 the four-toed salamanders. This interesting case of competing explanations between adaptation and other evolutionary factors research

---

[58] The presence of a "marked delay in, marked infrequency of, or absence of orgasm" is used in the definition of a sexual dysfunction, implying a "normal," non-dysfunctional, or functional, account of female orgasm when there is no single accepted account of female orgasm as noted above.

questions arises in the explanation of why so many miniaturized salamanders have four toes, as you can see in Figure 4.3 of the independent lineages.

Let us examine now how a methodological adaptationist would analyze this salamander example in Figure 4.3. We do not have to speculate: Methodological adaptationists Reeve and Sherman have written about this case, claiming that it may best be described as a case of natural selection, under the assumption that certain test conditions are undertaken. They propose that either: "(1) the production of four toes minimally disrupts the development of small individuals" or "(2) for small individuals, locomotion, clinging, and foraging, for example, are more efficient with four than with five toes" (1993, p. 22).

According to Reeve and Sherman, both hypotheses (1) and (2) are functional explanations, which they prefer to the developmental constraint and phyletic inertia explanation offered by Wake. They object that Wake has not explored either one of these selective explanations. They find arguments based on trait persistence to be "mysterious," focused as they themselves are on obtaining a function explanation. They propose what they call the "crucial" thought experiment: "What would be the evolutionary result if a mutant alternative trait arose and competed with the observed trait" (in this case, the four-toed form)? This thought experiment sets the trait in the context of selection rather than just development. This is, thus, for them, the "crucial" thought experiment

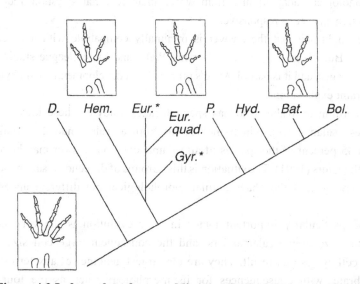

Figure 4.3 Independent four-toed lineages in the salamander family
Plethodontidae. Independent evolution of four-toed state in three separate lineages (Wake 1991, p. 548, Figure 1).

precisely because they have only one type of answer in mind, a function answer, one in the context of selection processes alone.

Let us take a look at this case using our logic of research questions framework. Slotting their proposed questions into the options for functional accounts, as they suggest, we get, as a set of answers to explain the convergence of multiple lineages onto a four-toed miniaturized state:

Box 4.6 METHODOLOGICAL ADAPTATIONIST RESEARCH QUESTION

What is the function of the four toes on the salamanders?

Possible and Responsive Answers

A: "The production of four toes minimally disrupts the development of small individuals."[59]
A: "For small individuals, locomotion, clinging, and foraging, for example, are more efficient with four than with five toes."[60]

Faulty and nonresponsive A: In contrast to functional answers, the developmental and phyletic constraint answer is "mysterious."[61]

But note that the first 'function' answer really is a developmental answer, contrary to the initial claim of Reeve and Sherman, and not fully a function answer after all. Because Reeve and Sherman have set up the problem using the methodological adaptationist framework, their research question requires a function answer as responsive.

We might say that the answer is minimally compatible with a function answer. But it's more, it's actually what Wake and his colleague studied in their lineages, and it is part of Wake's proposal of a developmental and phyletic constraint explanation.

According to Wake, who has spent decades studying these lineages of lungless salamanders, which include all tropical salamanders, a total of about 45 percent of the species of all salamanders, the answer may be a bit of both factors (1991). The situation is this. Groups of descendent salamanders are convergent if they have similar morphological but different ancestral states.

One particularly important factor in their evolution is the very large genome size of the salamanders, and the consequent large cell size and slow cell cycles that result. They are the largest genomes of any terrestrial vertebrate, with consequences for its morphology, such as the four-toe

---

[59] Reeve and Sherman 1993.   [60] Reeve and Sherman 1993.   [61] Reeve and Sherman 1993.

limitation with small size. It is a situation of indirect mechanisms of selection on body size: Large genome size is a selected mechanism for controlling body size via slow cell division times. Moreover, selection for body size has an incidental consequence of reduction of toe number. So, what looks like selection directly on toe number is likely actually selection on body size via slow cell division. And what seems to require adaptive explanations in cells (i.e., large genome size) is likely another mechanism for controlling body size. In sum, the connection is that large genome size results in slow cell division, slow cell division results in a small number of cells, which results in a small body size, and also because the slow process of cell division results in the last-to-form elements of body parts (like the fifth toe) not forming at all (Pers. Comm. J. R. Griesemer; Wake 1991).

Different branches of the salamander group independently have arrived at the four-toed state, but they are nearly identical in structure and are produced by the same complex of developmental options. As Wake writes, "[t]his is a direct example of design limitations, in which alternative states are sharply defined" (1991, p. 549). These are examples of how convergence can be a product of "design limitations in the form of developmental constraints, which are only indirectly related to adaptive processes." Hence, selection for very small size may have as an incidental side effect the reduction of a toe.

But Reeve and Sherman claim that all that Wake's developmental and phyletic constraints approach has to offer is a "description" but no "explanation."

As I understand scientific explanations, according to van Fraassen's "pragmatic theory of explanation" (1977), scientific explanations are answers to "why questions." As such, they provide the desired and sought-out information to answer a scientific inquiry as a description. Thus, *all scientific explanations are descriptions* in that they provide descriptions of systems or phenomena that answer the specific question sought in a scientific inquiry or question.

But Reeve and Sherman contrast descriptive and explanatory claims. See their dismissal of phyletic inertia as explanatory:

> ['phyletic' inertia] is sometimes treated as a unique mechanism of trait persistence. It is more appropriately regarded, however, as a *descriptive* label for the evolutionary stasis of a trait. (1993, p. 18; their emphasis)

This is much easier to see when we use the alternative evolutionary factors framework:

Comparing Boxes 4.6 and 4.7, the evolutionary factors approach makes very clear the complementary nature of the selective and developmental constraints explanations of the convergence of four-toedness in the salamanders.

> Box 4.7 Evolutionary Factors Research Question
>
> **What evolutionary factors contribute to the form and distribution of this trait?**
>
> *Possible and Responsive Answers:*
>
> A: Smaller salamanders are selected in certain environments, pushing miniaturization in many of the lineages (Selection)
> A: Four toed-ness is constrained through genome size and cell size once miniaturization of the lineage takes place (developmental and phyletic constraint).[62]
> A: Other selective and developmental factors can come into play in this context.

We don't have to take an approach like the methodological adaptationists where the views are seen as competing or mutually exclusive, as Reeve and Sherman do. And with the logic of research questions we can see precisely why they see some answers as real explanations and some as "mysterious" or mere descriptions.

Reeve and Sherman write:

> If structuralism is distinct from functionalism, it rests on unspecified – indeed mysterious – mechanisms of trait persistence . . . In summary, the structuralist approach either rests on a phenotype-set specification problem and therefore is fundamentally flawed or it dissolves into the recognition of selective developmental constraints, which is consistent with adaptationism and functionalism. (Reeve and Sherman 1993 p. 21)

For them, the only "real" explanations are answers to the function questions, that is, an answer is either "fundamentally flawed" or it is an explanation of adaptations.

Reeve and Sherman accuse the concept of a developmental constraint as being "almost as vague as that of phylogenetic inertia," which they savage in their article (1993, p. 20).

Everything else is apparently merely a description, under their view, as we can see with this salamander example. But Wake shows us that two answers – the selective and the developmental constraint – can be compatible and informative, as we saw in Box 4.7, in the salamanders.

4.3 Bicoid: Novelty and the Case of the Hox Genes

The logical problems with the adaptationist methodology can be seen clearly in the contrasts that we were just discussing. But we now turn to a case in genetics that is perhaps more difficult to analyze, one involving the evolution of a developmental gene of the famous and important *Hox* family, the genes

---

[62] See Wake (2009) for analysis of the first two evolutionary factors mentioned here in the salamander case working together.

responsible for placing segments and legs and heads in their correct places on the body forms of an organism.

The Hox gene we will look at is called *bicoid*, and it is descended from its ancestral gene, called *zen*, and they are both involved in determining where the head will appear on the fruitfly. The challenge is that *bicoid* has recently rapidly evolved from its ancestral gene, *zen*, and yet there seems to be no evidence for selection forces to account for such a change straightforwardly, directly on the *bicoid* gene (Carroll 2005).

One of the challenges for selection accounts of novel traits has to do with the dynamics of the selection process itself. Selection processes excel at refining traits to fit their local environments, but how do they come up with brand new traits? In fact, they rely on outside sources of variation to select upon, as we discussed in Section 2.5.

Moreover, one key thing to know in this example is that when natural selection acts on genes, it usually restricts variation in the gene, because it chooses the correct form of the gene and not the other forms (see Gannett 2010).

And I'm just using the "choosing" language here for convenience. The more strongly selection acts, the narrower the range of variation of the gene becomes. Likewise, the more relaxed the selective forces on a gene, the more variation accumulates in and surrounding the gene. When a gene is not expressed, it can vary more because selection is not acting to reduce the variation in the gene. As selective constraint is relaxed on a gene, natural variability in the gene is thereby increased, and we expect an increase proportional to the measure of the relaxation of selection on the gene. However, this may depend on the abundance of variation for a given trait. When selection is weak, different traits may vary at different rates. Thus one cannot simply infer the strength of selection from the amount of variation present. But we can make the reverse inference: Less selection means more variation in the bicoid gene.

In addition, while most genes are expressed by the complete zygote – the paired egg and sperm, which contains both the mother's and the father's cells – some genes are expressed only in the cells of the mother, called 'maternally expressed' genes (Barker et al. 2005).

Under maternally expressed genes, only half as much selection is taking place on the cell through the father's genes, which are not expressed at all, so they are twice as diverse as regular zygotic genes because of selection acting only on the mother's side. With zygotic selection, both the mother's and the father's genes are being selected, so we get narrowing of variability on both sides. But with maternally expressed genes, when only the mother's genes are being selected, we get increased variation of the father's genes, because they are not undergoing selection. We could say that the father's genes are "sheltered from selection forces" by the maternally expressed genes, because they are not being screened by natural selection. A selective death owing to genes in the zygote is twice as selective (it undergoes selection twice as strong) as a selective death of a zygote owing to genes in its mother.

With this increased diversity produced by maternally selected genes and the resultant expansion of variability on the father's genes, the genes can *evolve*, or *diverge between populations or species more quickly*, than the genes in the regular zygotic context, under which there is no advanced or accelerated variation. Such relaxed selection (on the father's genes) creates *more divergence* between populations and species than ordinary zygotic adaptive purifying selection does.

Despite the fact of increased genetic diversity because of maternally expressed genes and the accompanying relaxed selection, methodological adaptationists assume that *bicoid **must*** be adaptive in its evolution. Nevertheless, the increased genetic diversity arising from maternally expressed genes can explain their rapid evolution compared to the regular, zygotically expressed form, the ancestral *zen* form, as a live alternative to the adaptive account.

Michael Barker, Jeffrey Demuth, and Michael Wade (2005) performed an experiment on this type of maternally expressed selection that explained the evolution and selection of the *bicoid* and *zen* Hox genes in fruit flies. Population genetic theory predicts relaxed selection on maternally expressed genes with sex-limited expression, as we just discussed. They performed a series of experiments to investigate whether this phenomenon showed up in laboratory species of fruit fly, measuring variation and the expected accelerated evolution in the flies in the lab just as was predicted by theory.

Is any of this presentable using the terms we've been discussing so far? Can the methodological adaptationists handle this case? Does the evolutionary factors approach do better? Let us start by asking the methodological adaptationist question of the *bicoid* gene.

The Methodological Adaptationists' Research Question is: "What is the function of this trait?" In one sense, there is an obvious and obviously correct answer to this question put just so, which is: "the function is to put the head in place," as we get in Box 4.8.

An adaptationist might be completely satisfied with this answer, but usually, such an answer about function needs to be accompanied by an adaptive evolutionary account of the trait, saying how the trait was favored by selection through the evolutionary process, as shown in Box 4.9.

But is there such an account to be offered in this case? Again, one problem with selective accounts involving novel traits like *bicoid* here is that selection is really strong at providing adaptationist explanations about repurposing traits or revising

Box 4.8 Methodological Adaptationist Research Question

What is the function of this trait?

Possible and Responsive Answers

A: The function of this trait is to put the head in place.
A: The function of this trait is G.
Etc.

Box 4.9 Methodological Adaptationist Research Question: What Is the Function of This Trait?

Possible and Responsive Answers

A: The function of this trait is to put the head in place.
Full A: This trait has this function and correlation with fitness, and was selected through the evolutionary process, and we offer the following account to explain the history and selection of this trait.

them, but not so good at explaining *novel traits*, as we discussed in Section 2.5. In fact, questions concerning novel traits can rarely be satisfied with straightforward selective explanations, precisely because selection is usually about selecting, trimming, and revising something, and not about inducing something new. A significant part of the question about this trait concerns how such an important trait so recently appeared and rapidly changed relative to the rest of its genome. What is the evolutionary story behind the rapid evolution of this trait?

Does it help to turn to the evolutionary factors approach here?

What does the Barker, Demuth, and Wade story give us that goes beyond the function of the trait? They appeal to sex-limited selection in maternal effect genes having twice as much variation as zygotically expressed genes to account for the speedier evolution of the maternal-effect gene, *bicoid*, which controls the orientation and placement of the head in the body. Such an account cannot be arrived at through selection alone, as relaxed constraint on selection plays the key role in producing the increased variation in the maternally expressed genes. Such relaxed selection induces increased variation and novelty. Adaptation and natural selection work against novelty, hence, adaptive explanations can only explain the kind of novelty that results from refinement, and not this kind of novelty, which results from the relaxation of selective constraint.

Ultimately, this means that methodological adaptationist logic limits the types of traits that we can give adequate answers to, and forces us toward certain sorts of novelty and away from other sorts, those reliant on true increases

Box 4.10 Evolutionary Factors Research Question

What evolutionary factors account for the form and distribution of this trait?

Possible and Responsive Answers

A: Maternally expressed genes produce increased genetic variability through suppression of selection on the father's genes (easing of selection and increase in genetic variability).

A: Increased genetic variability produces wider genetic constraints controlling the field of exploration (constraints).

A: Wider constraints allow innovation and novelty in the *bicoid* gene (development and novelty, mutations).

in variation. Zygotic selection allows only winnowing selection on ancestral *zen* genes, that is, it allows repurposing only, as novelty. We must turn to the other evolutionary factors in order to account for the profound kind of novelty seen in this case, involving more than refinement of past cases. We are left needing to appeal to processes other than natural selection, which requires selecting from our list of other factors in the evolutionary factors approach. The adaptationist approach is limiting precisely by requiring the trait to be a repurposing only, otherwise the approach cannot fully explain it.

In other words, positive selection does not work as the key factor in the hard case that goes beyond the kind of novelty involving refinement of past traits, so we must turn to the other evolutionary factors of development and constraint, and as here, sex-limited traits arising from developmental factors. Constraints on selection are serving as a novelty-producing factor in this case. This is shown through predictions from population genetics and confirmation of those predictions through experimental data. In sum, this is an instance where methodological adaptationism's shortcomings are on display in a subtle but decisive manner, in showing how the significant factor of evolutionary novelty is not fully accounted for within that approach. Compare Boxes 4.9 and 4.10.

4.4 How Methodological Adaptationists Got Exaptation Wrong[63]

4.4.1 Background

There is one final type of evolutionary factor to consider before we close our consideration of adaptation. In 1982, Stephen Jay Gould and Elisabeth Vrba coined a new term, "exaptation," to mean a trait that contributes to fitness but

---

[63] "Exaptation Revisited: Changes Imposed by Evolutionary Psychologists and Behavioral Biologists," *Biological Theory* 12, pp. 50–65 (2017). E.A. Lloyd and S.J. Gould.

was not selected to do so. It was meant to contrast with "adaptation" (see Table 1.1). They also said that any trait that contributes to current fitness is an "aptation," which encompassed both adaptations and exaptations. The skull sutures in human babies (Figure 2.2) are a perfect example of an exaptation; they contribute to fitness, without having been built by selection for that purpose, that is, they are something with "vital current utility based on a cooptation of structures evolved in other contexts and for other purposes (or perhaps for no purpose at all)" (Gould 1991, p. 46). In other words, we have a trait that does not vary within the population, such as the existence of skull sutures in mammals, but which increases fitness over possible (but never actual) alternative traits, such as not having any skull sutures. If there had ever been significant variation in skull sutures, they would be selected against during birth, but there is not, so they are neither engineering nor selection-product adaptations. But they do contribute to fitness in mammals, in easing birth, so they are "aptations," with no history of selection, since baby skull sutures are just like reptile skull sutures, products of developmental constraint or architecture.

Exaptations became significant in a debate within evolutionary psychology in the early and mid-1990s, primarily because of an intervention by Gould, who claimed that many brain activities were likely to be exaptations – that is, effects that contribute to fitness, but were not selected to do so – rather than genuinely selected evolutionary functions in both the etiological and engineering sense.

This went against the orthodoxy of evolutionary psychology at the time, which was led by methodological adaptationists and was more inclined toward interpreting obviously fitness-increasing brain activities as evolutionarily selected functions.[64]

Unfortunately, in the context of that debate, Gould made an error of speaking and used the term "function" in the vernacular, or commonsense, manner, rather than in the technical, evolutionary engineering sense, as we shall see in a moment. Later on, some methodological adaptationists mistakenly took this occasion of Gould's misspeaking as confirmation that 'exaptation' had an evolutionarily 'designed' function and was a version of an engineering adaptation. This was a mistake because exaptation is not a version of adaptation; it is a distinct form of aptation. As a result, other methodological adaptationists provided a standard of evidence for exaptation that was inappropriate and based on an adaptationist

---

(This section of the Element, and the paper it is based on, were first written with Stephen Jay Gould as my co-author in the spring of 2002. All of the sections were drafted before he passed away in May 2002. It was our third co-written paper. I left it alone for over ten years, until I finally realized that he would have liked our arguments to be published, so I updated and completed the manuscript on my own. I would like to acknowledge Steve's contributions to this section, both conceptual and structural.)

[64] See, for example, Tooby and Cosmides (1997) debating Gould in the *New York Review of Books*.

worldview. These ideas have been rendered practically useless through their mistaken definitions and misapplications by some evolutionary psychologists over these last decades.

4.4.2 Evolutionary Function, a Misspoken Phrase, and a Misunderstanding

Various discussions of Gould and Vrba's (1982) concept of 'exaptation' have been plagued by terminological ambiguity. Specifically, varying uses of the concepts of 'current utility' and 'function' have contributed to uncertainty about when and how the notion of 'exaptation' should be used (see Table 1.1 for definitions). In this section, I shall focus on the issue of exaptation as it is discussed in two characteristic, widely cited, and very influential papers by David M. Buss, Martie G. Haselton, Todd K. Shackelford, April L. Bleske, and Jerome C. Wakefield (1998)[65], and Paul W. Andrews, Steven W. Gangestad, and Dan Matthews (2002a)[66].

In a widely cited paper aimed at explaining to psychologists the concepts of 'adaptation,' 'exaptation,' and 'spandrels,' David Buss and colleagues (1998) attempt to delineate the meanings and evidential requirements of these three terms. The notion of 'function' plays a central role in their analysis. In discussing adaptations, they write, "to propose that a trait is an adaptation is to propose that it was designed by natural selection in past environments to *serve some function*" (emphasis added; 1998, p. 85). Thus, they are using the same notions of function and adaptation we introduced in Section 1.

Now let us look at the way that Gould and Vrba used the term 'function' in their discussions of exaptation. The term 'exaptation' applies to two types of traits:

1. Traits that were adapted for one evolutionary function but were later co-opted (but *not* selected) to serve a different role; and
2. Traits that were correlates of growth, or accidental by-products, or "spandrels," that were later co-opted (but *not* selected) to serve another role.

Gould and Vrba also follow George Williams' terminology regarding function, stating that "the operation of an adaptation is its *function*,"[67] just as we find in Section 1. Similarly, the operation of a useful character not built by selection for its current evolutionary role is an "effect." Exaptations' effects do contribute to *current fitness*, by definition, but are not designed or selected to do so. "Adaptations have functions; *exaptations have effects*" (Gould and Vrba 1982, p. 6; emphasis added).

---

[65] Cited by 914 on Google Scholar, accessed June 2, 2020.
[66] Cited by 334 on Google Scholar, accessed June 2, 2020.
[67] Original emphasis; Gould and Vrba 1982, p. 6.

Gould and Vrba also suggest that the general phenomenon of being fit for a specific evolutionary role (current fitness) should be called 'aptation' rather than the usual 'adaptation'. This allows for a distinction between traits _selected_ for the current role, and those not. The most important part of these definitions is that exaptations must contribute to present fitness, but were not selected, historically, to do so. This is as opposed to adaptations that also contribute to present fitness, but which were naturally selected for their contributions to fitness and evolution of an adaptation _in the past_.

The regular requirements for adaptations demand that an adaptive account explain the adaptive feature(s) or modification(s) acquired through the selective process over time, its 'engineering' history of its complex and built-up adaptive traits. Under these definitions, both adaptations and secondary adaptations have selective histories, but exaptations do not. Recall that this definition of adaptation contrasts to the "product of selection" definition.

Note here that there is an important distinction to be made between secondary adaptations, which can be maintained by directional, stabilizing, or other forces of selection, and exaptations, which are not primarily maintained by selection. Stabilizing selection is therefore not the primary explanation of stability in the trait. Rather, exaptations maintain their status primarily through other biological mechanisms, such as developmental constraints, embryological considerations, or other structural means such as by-products or genetic correlations. The claim is: Selection is not the evolutionary cause that is maintaining the trait where it is.[68]

Significantly, exaptations can be modified by selection to serve yet a new evolutionary function: to become "secondary adaptations." For example, feathers originally served as exaptations for flight, under the assumption that they originally evolved for thermoregulation of the dinosaurs they appeared on. It is hypothesized that they were then utilized for short flights. However, the various subsequent modifications of feathers for flight are secondary adaptations, thus making certain features of the feathers engineering secondary adaptations themselves (Gould and Vrba 1982, p. 11). Contrary to a common objection, it is not the case that one generation of use in flight makes the feathers an engineering secondary adaptation, which requires generations of genetic change and distribution of that genetic change throughout the populations of the species. Only after such distributed genetic change and the acquisition of such genetic change and adaptive features by members of the species is the feather considered to be an engineering secondary adaptation for flight.

---

[68] E.g., Linde-Medina 2011; Newman & Bhat 2011; Amundson 2005.

Thus, if there is a stable trait in the population that contributes to fitness, on its face, it might be either an adaptation or an exaptation; it is simply an "aptation." It might have arrived at its present state through a selection process, in which case it is either a primary or a secondary adaptation. It may also have attained its current identity primarily through embryological or developmental constraints, and other structural features. The point is that these can explain why it is stable in the population, as an alternative and complement to stabilizing selection or other forces of selection. Relevant mechanisms might include genetic constraints and developmental details of the organism, correlations of growth, and by-products of other features of the organism. For example, the constancy of having seven cervical vertebrae in mammals, including mice, humans, and giraffes, has been revealed as a product of the involvement of Hox genes (related to the Bicoid case) and developmental constraints (Galis 1999).

Take the case of a novel use of a trait in a fitness-enhancing way. For example, an insect allele that was neutral before but in a new context provides protection against a novel pesticide would be, in the first exposure, not a selection-product or engineering adaptation for resistance until it undergoes natural selection (i.e., differential reproduction relying on this allele's protection). After selection, it would be considered a selection-product adaptation, as there is no new machinery or built-up trait involved in the process, but rather simple differential survival of the alleles (Lloyd 2001).

When Buss and colleagues confuse evolutionary "function" with exaptation's effects, they ignore this important point.

4.4.3 The Difference between Function and Effect

In a 1991 paper, Gould was not completely careful in separating "function" and "effect," and his misspeaking has led to a great deal of confusion. The occasion of the confusion was that Gould once described exaptation as "a feature, now useful to an organism, which did not arise as an adaptation for its present role, but was subsequently co-opted for its current *function*" (1991, p. 43, my emphasis). Here, Gould misspeaks and does not use "function" to refer strictly to the activity of an engineering adaptation. He instead uses "function" in a more vernacular sense so that it misleadingly appears to signify the useful action of a selected trait. Gould makes clear that such action involves enhancing current fitness (1991, p. 47).

This combination of statements led Buss and colleagues, due to their adaptationist biases, to conclude that an exaptation must have a "function" in the stricter, adaptive sense (1998, p. 539). Buss and colleagues end up confusing the requirement that exaptations contribute to current fitness with their own, added requirement that they serve a "function," understood as a solution to an *adaptive* problem. What is at stake in the confusion is the relative importance of using natural selection versus other evolutionary factors to account for currently useful, 'aptive,' fitness-enhancing traits, and in general, the overall power of adaptive explanations.

They contrast selective *functions*, which for Buss et al. include exaptations, with "biologically functionless uses," which they refer to as "effects," "consequences," or "byproducts" (1998, p. 539). Thus, they have severely altered Gould and Vrba's terminology when they *exclude "effects"* from the performance of exaptations. These effects are precisely what Gould and Vrba defined as the outcomes of exaptations! What Buss and colleagues have described here are, in fact, secondary adaptations, not the original nonadaptations that signify "exaptations."

4.4.4 Applying the Logic of Research Questions

Let's look at this using the Logic of Research Questions.

As we can see, Buss and colleagues have mistakenly turned a nonresponsive answer, the true exaptationist position, into a responsive answer that satisfies their function research question (See Box 4.11, below). As such, the actual definition of 'exaptation' is missing and is not being actively considered by the evolutionary psychologists under their research question in a paper that has been cited over 900 times.

The alternative, "Evolutionary Factors" research question, offers the full variety of evolutionary explanations, including adaptation through selection, exaptation through cooptation (under the original definition), and so on. These questions and a few of their answers can be seen in Box 4.12.

Box 4.11 METHODOLOGICAL ADAPTATIONIST RESEARCH QUESTION

"What is the function of this trait?"

Possible and Responsive Answers

A: This trait has the function F, an adaptation.
A: This trait has the function G . . .
A: *This trait has the function H as an exaptation, as a result of a process of selection* (*Under Buss and colleagues' *mistaken* assumption that exaptations have functions, and are products of selection)
Missing Nonresponsive A: This trait does not have a function, and is an exaptation, as a result of a process of co-optation under Gould & Vrba's definition (i.e., nonresponsive to the function question, under the *true* definition of exaptation, where exaptations do not have functions, but rather only effects, a "null" result according to the adaptationists).
Nonresponsive A: This trait is a by-product of selection on another trait (nonresponsive to the function question, a "null" result according to the adaptationists).

Box 4.12 Evolutionary Factors Research Question

"What evolutionary factors account for the form and distribution of this trait," for example, or, "Does this trait have a function?"

Possible Answers:

A: This trait occurs in the population because it has the function F (i.e., the trait is an engineering adaptation or secondary adaptation).

. . .

. . .

A: This trait has its current form and distribution because it is an exaptation as a result of co-optation but *not selection* (cannot be combined with the first two adaptation answers, unless the traits are different).

A: This trait has its current form and distribution because of some combination of the above factors.

Etc.

Under the evolutionary factors approach, exaptation is one possible answer to the question "What evolutionary factors account for the form and distribution of this trait?", and it appeals to a process of cooptation *in contrast to* a process of selection.

Buss and colleagues misleadingly discuss the sources or mechanisms of co-optations of structures, under their faulty definition, discussed above. In considering birds' feathers, which were originally evolved for thermoregulation but subsequently co-opted for flight, they write "it is clearly natural selection that is responsible for transforming an existing structure into a new, modified structure with a different function" (1998, p. 541). But Buss and colleagues have conflated two processes here. They've taken one process in which feathers, without physiological modification, were co-opted for flight through use (an exaptation), and confused it with another in which feathers were subsequently adaptively modified for enhanced flight capability (a secondary engineering adaptation, which has a function in both the etiological and design sense).

They object to Gould's use of some mechanisms of co-optation, like cognitive capacities and motivational mechanisms, because they are clearly not cases of natural selection. They are at pains to distinguish current uses that are not *"co-opted by natural selection"* for their current uses from those that are naturally selected. But that is the whole point of introducing "exaptation." Buss and colleagues fail to see that Gould does not require that the trait be "co-opted" *by natural selection* (which really means "selected," based on their use of the term) in order to be an exaptation; this is a result of their own collapse of the two processes of exaptation and secondary adaptation.

Some measure of the responsibility for this confusion is Gould's. His single misspeaking involving the term "function" in the 1991 paper was seized on by Buss and colleagues, and led them to search for an original function and a "distinct later function" in secondary adaptations, their "exaptations" (1998, p. 541). When finding no later adaptive function, Buss and colleagues conclude that something is amiss. Moreover, when they understand exaptations as arising from earlier functions, they are missing half of the definition of exaptation, in which many traits with current uses are derived from traits with no adaptive function at all (i.e., spandrels).

But what turns on this confusion? The answer becomes clear in Buss and colleagues' discussion of the evidence required to show that something is an exaptation. They require specification of (1) the original adaptation or by-product that was later coopted for the exaptation; (2) the causal mechanism responsible for the co-opting (e.g., on their account, natural selection or a selective motivational mechanism; note that they do not include any devel-opmental constraints or mechanisms or nonselective factors); and (3) "the exapted biological function" of the trait, "that is, the manner in which it contributes to the solution to an adaptive problem of survival or reproduction" (1998, p. 542).

Here, Buss and colleagues' notion that an exaptation must have a biological function is doing some real work. As they put it, both the concepts of exapta-tion and adaptation invoke function, "therefore, both must meet the concep-tual and evidentiary standards for invoking function." It becomes clear that these standards are much too strict to bear the weight of what Gould and Vrba have defined as exaptations. Buss and colleagues write, for example, that the criteria for a proposed function include "the hallmarks of special design, including specialization of function, for solving a particular adaptive prob-lem" (1998, p. 546). Thus, it seems that Buss and colleagues are using Williams' engineering definition of function and are mistakenly assuming that exaptations must exhibit this kind of function, whereas exaptations are actually demarcated precisely to avoid such a burden.

Buss and colleagues claim that selectionist explanations "would generally be necessary for explaining how functionless by-products are transformed into coopted spandrels that perform specific functions," thus making complete their mistaken view of the connection between function and co-optation. In an adaptationist spin, all of this is used to defend their sought-for conclusion that natural selection is *"the basic explanatory principle in biology"* (1998, p. 543; emphasis added).

The key problem is that their view does not allow for exaptation to serve its original role in evolution, as a potential alternative evolutionary factor, since it has been defined away as a selective scenario, often involving stabilizing selection. Thus, the differences between the evolutionary psychologists Buss and colleagues, and Vrba and Gould are clear, as we can see from the Logic of Research Questions.

It is badly mistaken to read exaptations as examples of secondary adaptations subject to stabilizing selection. Because methodological adaptationists, like Buss and colleagues, ask the "function" research question, their question needs a function answer, thus producing a function-based definition of exaptation. This is incorrect. In sum, Gould's single, misspoken mention of the '*function*' of an exaptation in his 1991 paper has enabled Buss and colleagues to completely reverse the intended meaning of 'exaptation.'

A similar problem arises with evolutionary psychologists Andrews et al. (2002a, 2002b). In a paper in which they claim to define exaptation and propose definitive standards of evidence for adaptation and exaptation, they assume adaptations and functions for traits at the beginning of inquiry, proposing that all adaptive accounts must be eliminated as alternatives before a nonadaptive account may be seriously considered (Andrews et al. 2002a p. 534; 2002b p. 504). This approach suffers from similar faults as Buss et al.'s, but we won't detail it here; analysis can be found in Lloyd and Gould (2017).

4.5 Conclusion

Before we end this discussion of the risks or shortcomings of methodological adaptationism, I must acknowledge that many will object to my critique of adaptationism: "But methodological adaptationism is so useful! Surely you are not advocating sacrificing our most fruitful research tool?!"

And no, I am *not* doing so, since the evolutionary factors framework includes the use of adaptation, and the search for connections to fitness, function, and "design" may even serve as a first "go-to" algorithm.

The evolutionary factors framework may advocate starting[69] research by trying out the adaptive answer to "Does this trait have a function?" and learning and keeping at the top of the mind as real causal alternatives the other evolutionary factors. (Sometimes this different set of answers may require a different laboratory setup or tools, e.g., David Wake's work on salamanders.[70]) That is the difference between methodological

---

[69] Of course, different researchers may start with different questions. My point is that typical researchers in animal behavior, evolutionary psychology, etc., will likely want to start by searching for adaptations of a given trait, and they can still do so under the evolutionary factors approach.

[70] Wake 1991, 2009; Griesemer 2013, 2015; Wagner 2000.

adaptationism and the evolutionary factors framework: The nonadaptive explanations are real, causal, live alternatives that can be supported with evidence. Moreover, this evidential support for the full variety of evolutionary factors must be recognized when it is presented, and not treated within a mutually exclusive framework of evidence.

To reiterate, I am not in any way against adaptive explanations themselves. But I have highlighted some *risks* of a very popular approach to research into evolutionary causes. These dangers become obvious when we examine the logic of the research questions and their relevant answers, within the methodological adaptationist approach and the contrasting evolutionary factors framework. When a research approach makes any particular types of hypothesis especially difficult to entertain or accept, it deserves serious scrutiny. Keeping the logic of the research questions in mind when dealing with the scientific errors committed by adaptationists allows us to analyze and explain them straightforwardly. The presence of researchers like Symons, Wade, or Wake, who engaged in their research using the more inclusive evolutionary factors methodology, exemplifies a living available alternative method.

Evolutionary adaptations are fascinating aspects of living beings. Understanding not just whether to study them, but also **how** to study them, is crucially important to the success of evolutionary science.[71]

---

[71] **Suggested Further Reading:** For evidence for by-product/bonus account of female orgasm, see Lloyd 2005a and Levin 2014; for the hereditary basis of orgasm, see Dawood et al. 2005 and Dunn et al. 2005; for total lack of any genetic correlation with fitness, see Zietsche and Santtila 2011, 2013. See Marlene Zuk (2006) for a review of my 2005 book; see Dixson 2012 for the authoritative study of primate sexuality, including comparative study of human beings; see Wake 1991, 2009 for more on the salamander case. See Barker et al. 2005 for more on the Bicoid, Zen, and Hox genes.

References

Ackerly, D. D. 2004. "Adaptation, niche conservatism, and convergence: Comparative studies of leaf evolution in the California chaparral." *The American Naturalist* 163(5): 654–71. https://doi.org/10.1086/383062

"Adaptive evolution | boundless biology." n.d. Accessed November 28, 2018. https://courses.lumenlearning.com/boundless-biology/chapter/adaptive-evolution/.

Airoldi, G., 2018. "More than fitness. A robustness-based proposal of a logical space to classify processes behind evolutionary phenomena." *Kairos. Journal of Philosophy & Science* 20(1): 89–112.

Alcock, J. 1987. "Ardent adaptationism." *Natural History* 96(4): 4.

Alcock, J. 1998. "Unpunctuated equilibrium in the natural history essays of Stephen Jay Gould." *Evolution and Human Behavior* 19(5): 321–36. https://doi.org/10.1016/S1090-5138(98)00029-4.

Allen, C. 2002. "Real traits, real functions?" In Ariew, A., R. Cummins, and M. Perlman (eds.) *Functions: new essays in the philosophy of psychology and biology.* New York: Oxford University Press. 373–89.

American Psychiatric Association. 2013. "Sexual dysfunctions" in *Diagnostic and statistical manual of mental disorders* (5th ed.). Washington, DC. https://doi.org/10.1176/appi.books.9780890425596.dsm13

Amundson, R. 1994. "Two concepts of constraint: Adaptationism and the challenge from developmental biology." *Philosophy of Science* 61(4): 556–78. https://doi.org/10.1086/289822

Amundson, R.. 1998. "Typology reconsidered: Two doctrines on the history of evolutionary biology." *Biology and Philosophy* 13(2): 153–77. https://doi.org/10.1023/A:1006599002775

Amundson, R. 2001. "Adaptation and development: On the lack of common ground." In Orzack, S., and E. Sober (eds.). *Adaptationism and optimality.* Cambridge University Press, New York. 303–34.

Amundson, R. 2005. *The changing role of the embryo in evolutionary thought: Roots of Evo-Devo.* Cambridge University Press.

Andrews, P. W., S. W. Gangestad, and D. Matthews. 2002a. "Adaptationism, exaptationism, and evolutionary behavioral science." *Behavioral and Brain Sciences* 25(4): 534–47. https://doi.org/10.1017/S0140525X02530090

Andrews, P. W., S. W. Gangestad, and D. Matthews. 2002b. "Adaptationism – how to carry out an exaptationist program." *The Behavioral and Brain Sciences* 25(4): 489–504.

Aoki, K. 1984. "A population genetic model of the evolution of oblique cultural transmission." *Proceedings of the Japan Academy, Series B* 60(8): 310–13

Aoki, K. 1986. "A stochastic model of gene-culture coevolution suggested by the 'culture historical hypothesis' for the evolution of adult lactose absorption in humans." *Proceedings of the National Academy of Sciences* 83(9): 2929–33.

Arnold, S. J. 1994. "Is there a unifying concept of sexual selection that applies to both plants and animals?" *The American Naturalist* 144: S1–12. https://doi.org/10.1086/285650

Baker, R. R., & Bellis, M. A. 1993. "Human sperm competition: ejaculate manipulation by females and a function for the female orgasm." *Animal Behaviour* 46(5): 887–909.

Barash, D. 2005. "Let a thousand orgasms bloom [Book Review of *The case of the female orgasm*. E. A. Lloyd]." *Evolutionary Psychology* 3(1): 347–54. https://doi.org/10.1177/147470490500300123

Barash, D. P., and J. E. Lipton. 2009. *How women got their curves and other just-so stories: evolutionary enigmas*. Columbia University Press.

Barker, M. S., J. P. Demuth, and M. J. Wade. 2005. "Maternal expression relaxes constraint on innovation of the anterior determinant, bicoid." *PLoS Genetics* 1(5): e57. https://doi.org/10.1371/journal.pgen.0010057

Beatty, J. 1987. "Natural selection and the null hypothesis." In J. Dupre (ed.) *The latest on the best: essays on evolution and optimality*. 53–75. The Massachusetts Institute of Technology Press.

Boyd, R., and P. J. Richerson. 2009. "Culture and the evolution of human cooperation." *Philosophical Transactions of the Royal Society of London B: Biological Sciences* 364(1533): 3281–88. https://doi.org/10.1098/rstb.2009.0134.

Buss, D. M., M. G. Haselton, T. K. Shackelford, A. L. Bleske, and J. C. Wakefield. 1998. "Adaptations, exaptations, and spandrels." *American Psychologist* 53(5): 533–48. https://doi.org/10.1037/0003-066X.53.5.533

Carroll, S. B. 2005. *Endless forms most beautiful: the new science of evo devo and the making of the animal kingdom*. W. W. Norton.

Cavalli-Sforza, L., and M. W. Feldman. 1973. "Models for cultural inheritance. I. Group mean and within group variation." *Theoretical Population Biology* 4(1): 42–55.

Cavalli-Sforza, L. L., and Feldman, M. W. 1981. *Cultural transmission and evolution: A quantitative approach*. Princeton University Press.

Charnov, E. L., and G. A. Parker. 1995. "Dimensionless invariants from foraging theory's marginal value theorem." *Proceedings of the National Academy of Sciences* 92(5): 1446–50. https://doi.org/10.1073/pnas.92.5.1446

Cheng, H., and W. M. Muir. 2005. "The effects of genetic selection for survivability and productivity on chicken physiological homeostasis." *World's Poultry Science Journal* 61(3): 383–97. https://doi.org/10.1079/WPS200464

Chevalier-Skolnikoff, S. 1974. "Male-female, female-female, and male-male sexual behavior in the stumptail monkey, with special attention to the female orgasm." *Archives of Sexual Behavior* 3(2). 95–116.

Cosmides, L., and J. Tooby. 1994. "Better than rational: evolutionary psychology and the invisible hand." *The American Economic Review* 84(2): 327–32.

"Cranial sutures and fontanels." n.d. Mayo Clinic. Accessed December 5, 2018. www.mayoclinic.org/diseases-conditions/craniosynostosis/multimedia/cranial-sutures-and-fontanels/img-20006785

Cummins, R. 1975. "Functional analysis." *The Journal of Philosophy* 72(20): 741–65. doi:10.2307/2024640

Darwin, C. 1859. *On the origin of species by means of natural selection, or preservation of favoured races in the struggle for life.* London: John Murray.

Darwin, C. 1871. *The descent of man and selection in relation to sex.* London: John Murray.

Dawood, K., K. M. Kirk, M. J. Bailey, P. W. Andrews, and N. G. Martin. 2005. "Genetic and environmental influences on the frequency of orgasm in women." *Twin Research and Human Genetics* 8(1): 27–33. https://doi.org/10.1375/twin.8.1.27.

Delph, L. F., and T. Ashman. 2006. "Trait selection in flowering plants: How does sexual selection contribute?" *Integrative and Comparative Biology* 46(4): 465–72. https://doi.org/10.1093/icb/icj038.

Delph, L. F., and C. R. Herlihy. 2012. "Sexual, fecundity, and viability selection on flower size and number in a sexually dimorphic plant." *Evolution: International Journal of Organic Evolution* 66(4): 1154–66. https://doi.org/10.1111/j.1558-5646.2011.01510.x

Dewsbury, D. A. 1992. "Essay on contemporary issues in ethology: on the problems studied in ethology, comparative psychology, and animal behavior." *Ethology* 92(2): 89–107. https://doi.org/10.1111/j.1439-0310.1992.tb00951.x

"Digital morphology at the University of Texas." n.d. Accessed December 5, 2018. http://digimorph.org/index.phtml

Dixson, A. 2009. *Sexual selection and the origins of human mating systems.* Oxford University Press.

Dixson, A. 2012. *Primate sexuality.* 2nd ed. Oxford University Press.

Douglas, A. E., and J. H. Werren. 2016. "Holes in the hologenome: Why host-microbe symbioses are not holobionts." *MBio*. 7(2).10.1128/mBio.02099–15

Dunn, K. M., L. F. Cherkas, and T. D. Spector. 2005. "Genetic influences on variation in female orgasmic function: a twin study." *Biology Letters* 1(3): 260–63. https://doi.org/10.1098/rsbl.2005.0308

El-Sabaawi, R. W., E. Zandonà, T. J. Kohler, M. C. Marshall, J. M. Moslemi, J. Travis, A. López-Sepulcre, R. Ferriere, C. M. Pringle, S. A. Thomas, and D. N. Reznick. 2012. "Widespread intraspecific organismal stoichiometry among populations of the trinidadian guppy." *Functional Ecology* 26(3): 666–76. https://doi.org/10.1111/j.1365-2435.2012.01974.x

Endler, J. A. 1978. "A predator's view of animal color patterns." In Hecht M. K., W. C. Steere, and B. Wallace (eds.), *Evolutionary Biology. Evolutionary Biology, vol II*. Springer, Boston. https://doi.org/10.1007/978-1-4615-6956-5_5

Endler, J. A. 1986. *Natural selection in the wild*. Princeton University Press.

Fairbanks, D. J., and B. Rytting. 2001. "Mendelian controversies: a botanical and historical review." *American Journal of Botany* 88(5): 737–52.

Feldman, M. W., and K. N. Laland. 1996. "Gene-culture coevolutionary theory." *Trends in Ecology & Evolution* 11(11): 453–57.

Feldman, M. W., and L. Cavalli-Sforza. 1985. *On the theory of evolution under genetic and cultural transmission with application to the lactose absorption problem*. Stanford Institute for Population and Resource Studies.

Fisher, S. 1973. *Understanding the female orgasm*. New York: Basic Books

Fitelson, B., 2001. "A Bayesian account of independent evidence with applications." *Philosophy of Science* 68(S3): S123–S140.

Freud, S. [1905] 1953. "Three essays on the theory of sexuality." In *The standard edition of the complete psychological works of Sigmund Freud, volume VII (1901–1905): A case of hysteria, three essays on sexuality and other works*. 123–246.

Freud, S. [1923] 1963. "Freud: introductory lectures on psychoanalysis." *The standard edition of the complete psychological works of Sigmund Freud*.

Futuyma, D. and M. Kirkpatrick. 2017. *Evolution* (4th ed.). Sinauer Oxford University Press.

Galis, F. 1999. "Why do almost all mammals have seven cervical vertebrae? Developmental constraints, hox genes, and cancer." *Journal of Experimental Zoology* 285(1): 19–26.

Gannett, L. 2010. "Questions asked and unasked: how by worrying less about the 'really real' philosophers of science might better contribute to debates about genetics and race." *Synthese* 177(3): 363–85. https://doi.org/10.1007/s11229-010-9788-1

Geary, D. C., and M. V. Flinn. 2001. "Evolution of human parental behavior and the human family." *Parenting* 1(1–2): 5–61. https://doi.org/10.1080/15295192.2001.9681209.

Glymour, C. 1980. *Theory and evidence*. Princeton University Press.

Godfrey-Smith, P. 2001. "Three kinds of adaptationism." In Orzack, S. and E. Sober (eds.). *Adaptationism and optimality*. 344–62. Cambridge University Press.

Gould, S. J. 1987a. "Freudian slip." *Natural History* 96(2): 14–21.

Gould, S. J. 1987b. "Reply to Alcock." *Natural History* 96(4): 4–6.

Gould, S. J. 1991. "Exaptation: a crucial tool for an evolutionary psychology." *Journal of Social Issues* 47(3): 43–65. https://doi.org/10.1111/j.1540-4560.1991.tb01822.x

Gould, S. J. 2002. *The structure of evolutionary theory*. Harvard University Press.

Gould, S. J., and E. S. Vrba. 1982. "Exaptation – a missing term in the science of form." *Paleobiology* 8(1): 4–15. https://doi.org/10.1017/S0094837300004310

Gould, S. J., and R. C. Lewontin. 1979. "The spandrels of San Marco and the panglossian paradigm: a critique of the adaptationist programme." *Proceedings of the Royal Society of London. Series B Biological Science* 205(1161): 581–98. https://doi.org/10.1098/rspb.1979.0086

Griesemer, J. 2013. "Integration of approaches in David Wake's model-taxon research platform for evolutionary morphology." *Studies in History and Philosophy of Science Part C: Studies in History and Philosophy of Biological and Biomedical Sciences* 44(4, Part A): 525–36. https://doi.org/10.1016/j.shpsc.2013.03.021

Griesemer, J. 2015. "What salamander biologists have taught us about evo-devo." In Love, A. C. (ed.) *Conceptual change in biology: Scientific and philosophical perspectives on evolution and development*. 271–301. Boston Studies in the Philosophy and History of Science. Springer. https://doi.org/10.1007/978-94-017-9412-1_13.

Griffiths, A. J. F., Wessler, R. C. Lewontin, W. M. Gelbart, D. T. Suzuki, and J. H. Miller. 2005. *An introduction to genetic analysis* (8th ed.). Macmillan.

Griffiths, P. E. 1996. "The historical turn in the study of adaptation." *The British Journal for the Philosophy of Science* 47(4): 511–32. https://doi.org/10.1093/bjps/47.4.511

Harrison, D. F. N. 1980. "Biomechanics of the giraffe larynx and trachea." *Acta Oto-Laryngologica* 89 (3–6): 258–64. https://doi.org/10.3109/00016488009127136.

Harrison, D. F. N. 1981. "Fibre size frequency in the recurrent laryngeal nerves of man and giraffe." *Acta Oto-Laryngologica* 91(1–6): 383–9. https://doi.org/10.3109/00016488109138519

Hite, S. 1976. *The Hite report on female sexuality.* Bertelsmann.

Hosken, D. J. 2008. "Clitoral variation says nothing about female orgasm." *Evolution & Development* 10(4): 393–95. https://doi.org/10.1111/j.1525 -142X.2008.00247.x

Hrdy, S. B. 2005. "Cooperative breeders with an ace in the hole," In Voland E., A. Chasiotis, and W. Schiefenhövel (eds.) *Grandmotherhood: The evolutionary significance of the second half of female life.* 295–317.

"Inkscape tutorial: tracing bitmaps | Inkscape." n.d. Accessed November 29, 2018. https://inkscape.org/doc/tutorials/tracing/tutorial-tracing.html

Inside Nature's Giants | PBS. n.d. Accessed November 29, 2018. www.pbs.org /show/inside-natures-giants/

Jablonski, D. 2008. "Species selection: Theory and data." *Annual Review of Ecology, Evolution, and Systematics* 39(1): 501–24. https://doi.org/10.1146 /annurev.ecolsys.39.110707.173510

Jablonski, D., and G. Hunt. 2006. "Larval ecology, geographic range, and species survivorship in Cretaceous mollusks: Organismic versus species-level explanations." *The American Naturalist* 168(4): 556–64. https://doi .org/10.1086/507994

Jacob, F. 1977. Evolution and tinkering. *Science* 196(4295): 1161–66.

Kerr, B., and P. Godfrey-Smith. 2002. "Individualist and multi-level perspectives on selection in structured populations." *Biology and Philosophy* 17(4): 477–517.

Ketterson, E. D., and V. Nolan Jr. 1999. "Adaptation, exaptation, and constraint: A hormonal perspective." *The American Naturalist* 154 (S1): S4–25. https:// doi.org/10.1086/303280.

Kinsey, A. C., W. B. Pomeroy, C. E. Martin, and P. H. Gebhard. 1953. *Sexual behavior in the human female.* Indiana University Press.

Komisaruk, B., C. Beyer-Flores, and B. Whipple. 2006. *The science of orgasm.* Baltimore: John Hopkins University Press.

Komisaruk, B. R., N. Wise, E. Frangos, W. Liu, K. Allen, and S. Brody. 2011. "Women's clitoris, vagina, and cervix mapped on the sensory cortex: FMRI evidence." *The Journal of Sexual Medicine* 8(10): 2822–30. https://doi.org /10.1111/j.1743-6109.2011.02388.x

Laland, K. N., J. Odling-Smee, and M. W. Feldman. 2000. "Niche construction, biological evolution, and cultural change." *Behavioral and Brain Sciences* 23(1): 131–46. https://doi.org/10.1017/S0140525X00002417.

Lemmon, W. B., and M. L. Allen. 1978. "Continual sexual receptivity in the female chimpanzee (Pan troglodytes)." *Folia Primatologica* 30, no. 1: 80–88.

Levin, R. J. 2011. "Can the controversy about the putative role of the human female orgasm in sperm transport be settled with our current physiological knowledge of coitus?" *The Journal of Sexual Medicine* 8(6): 1566–78. https://doi.org/10.1111/j.1743-6109.2010.02162.x.

Levin, R. J. 2014. "Should the clitoris become a vestigial organ by personal psychological clitoridectomy? A critical examination of the literature." *Journal of Womens Health, Issues and Care.* https://doi.org/10.4172/2325-9795.1000159

Levin, R. J. 2015. "Recreation and procreation: A critical view of sex in the human female." *Clinical Anatomy* 28(3): 339–54. https://doi.org/10.1002/ca.22495.

Levins, R. 1966. "The strategy of model building in population biology." *American Scientist* 54(4): 421–31.

Lewens, T. 2009. "Seven types of adaptationism." *Biology & Philosophy* 24 (2): 161–82 https://doi.org/10.1007/s10539-008-9145-7.

Lewontin, R. C. 1978. "Adaptation." *Scientific American* 239(3): 213–31.

Lewontin, R. C. 1983. "Gene, Organism and Environment." In Bendal, D. S. (ed.) *Evolution from molecules to men.* Cambridge University Press.

Lewontin, R. C., and L. C. Dunn. 1960. "The evolutionary dynamics of a polymorphism in the house mouse." *Genetics* 45(6): 705–22.

Li, D., and R. O'Loughlin. Manuscript. "Tractability assumptions, holism, and model robustness." Submitted.

Linde-Medina, M. 2011. "Adaptation or exaptation? The case of the human hand." *Journal of Biosciences* 36(4): 575–85. https://doi.org/10.1007/s12038-011-9102-5

Linquist, S. P. 2006. "Sometimes an orgasm is just an orgasm." *Metascience* 15 (2): 411–19.

Lloyd, E. A. 1983. "The nature of Darwin's support for the theory of natural selection." *Philosophy of Science* 50 (1): 112–29. https://doi.org/10.1086/289093

Lloyd, E. A. 1988/1994. *The structure and confirmation of evolutionary theory.* Princeton University Press.

Lloyd, E. A. 2001. "Units and levels of selection: An anatomy of the units of selection debates." In Rama S., C. B. Krimbas, J. Beatty, and D. B. Paul (eds.). *Thinking about evolution: Historical, philosophical, and political perspectives* (2): 267–91. Cambridge University Press.

Lloyd, E. A. 2005a. *The case of the female orgasm: Bias in the science of evolution.* Cambridge, MA: Harvard University Press.

Lloyd, E. A. 2005b. "Why the gene will not return." *Philosophy of Science* 72(2). 287–310.

Lloyd, E. A. 2013. "Stephen J. Gould and adaptation: San Marco 33 years later." In *Stephen J. Gould: The Scientific Legacy.* Springer, Milano. 21–35

Lloyd, E. A. 2015a. "Adaptationism and the logic of research questions: How to think clearly about evolutionary causes." *Biological Theory* 10(4): 343–62. https://doi.org/10.1007/s13752-015-0214-2.

Lloyd, E. A. 2015b. "Model robustness as a confirmatory virtue: The case of climate science." *Studies in History and Philosophy of Science Part A* 49: 58–68.

Lloyd, E. A. 2017. "Units and levels of selection," *Stanford Encyclopedia of Philosophy*. www.plato.stanford.edu/entries/selection-units/

Lloyd, E. A., and S. J. Gould. 1993. "Species selection on variability." *Proceedings of the National Academy of Sciences* 90(2): 595–99. https://doi .org/10.1073/pnas.90.2.595

Lloyd, E. A., and S. J. Gould. 2017. "Exaptation revisited: Changes imposed by evolutionary psychologists and behavioral biologists." *Biological Theory* 12(1): 50–65. https://doi.org/10.1007/s13752-016-0258-y

Lloyd, E. A., R. C. Lewontin, and M. W. Feldman. 2008. "The generational cycle of state spaces and adequate genetical representation." *Philosophy of Science* 75(2), 140–56.

Lloyd, E. A., and M. J. Wade. 2019. "Criteria for holobionts from community genetics." *Biological Theory* 14(3): 151–70. 10.1007/s13752-019–00322-w

Lloyd, E. A., D. S. Wilson, and E. Sober. 2011. "Evolutionary mismatch and what to do about it." *The Evolution Institute* (blog). November 13, 2012. https://evolution-institute.org/evolutionary-mismatch-and-what-to-do-about -it/

Losos, J. B. 2011. "Seeing the forest for the trees: The limitations of phylogenies in comparative biology." *The American Naturalist* 177(6): 709–27. https://doi.org/10.1086/660020

Love, A. C. 2003. "Evolutionary morphology, innovation, and the synthesis of evolutionary and developmental biology." *Biology and Philosophy* 18(2): 309–45.

Love, A. C. 2006. "Evolutionary morphology and evo-devo: Hierarchy and novelty." *Theory in Biosciences* 124(3): 317–33.

Marshall Cavendish Corporation. 2010. *Mammal anatomy: An illustrated guide. Illustrated edition*. New York: Cavendish Square.

Martins, E. P. 2000. "Adaptation and the Comparative Method." *Trends in Ecology & Evolution* 15(7): 296–99. https://doi.org/10.1016/S0169-5347(00)01880-2.

Masters, W. H. and V. E. Johnson. 1965. "The sexual response of the human female." In Money, J. (ed.). *Sex Research: New Developments*. 53–112. New York: Holt, Rinehart, and Winston.

Masters, W. H. and V. E. Johnson. 1966. *The human sexual response. Boston*: Little, Brown.

Maynard Smith, J. 1978. "Models in ecology." CUP Archive.

Maynard Smith, J. 1993. *The theory of evolution.* Cambridge University Press.

Mayr, E. 1983. "How to carry out the adaptationist program?" *The American Naturalist* 121(3): 324–34. https://doi.org/10.1086/284064

Medeyko, V. 1986. *The Collins Encyclopedia of Animal Evolution.* Berry, R. J. and A. Hallam (eds.). London: Collins.

Millstein, R. L. 2008. "Distinguishing drift and selection empirically: 'The great snail debate' of the 1950s." *Journal of the History of Biology* 41(2): 339–67. https://doi.org/10.1007/s10739-007-9145-5.

Muller, G. B., & Wagner, G. P. 1991. "Novelty in evolution: restructuring the concept." *Annual Review of Ecology and Systematics* 22(1): 229–56.

Murphy, C. G. 1998. "Interaction-independent sexual election and the mechanisms of sexual selection." *Evolution* 52 (1): 8–18. https://doi.org/10.1111/j .1558-5646.1998.tb05133.x

Newman, S. A. 1988. "Lineage and pattern in the developing vertebrate limb." *Trends in Genetics* 4 (12): 329–32.

Newman, S. A., and R. Bhat. 2008. "Dynamical patterning modules: Physico-genetic determinants of morphological development and evolution." *Physical Biology* 5 (1): 015008. https://doi.org/10.1088/1478-3975/5/1/015008

Newman, S. A., and R. Bhat. 2011. "Lamarck's dangerous idea." In Jablonka E. and S. B. Gissis (eds.), *Transformations of Lamarckism: From subtle fluids to molecular biology.* 157–70. Cambridge, Massachusetts Institute of Technology Press.

Nijhout, F. H. 1991. *The development and evolution of butterfly wing patterns.* Smithsonian Institution Scholarly Press.

Oakley, S. H., C. M. Vaccaro, C. C. Crisp, M. V. Estanol, A. N. Fellner, S. D. Kleeman, and R. N. Pauls. 2014. "Clitoral size and location in relation to sexual function using pelvic MRI." *The Journal of Sexual Medicine* 11(4): 1013–22. https://doi.org/10.1111/jsm.12450.

Odenbaugh, J. 2018. "Emotions." Ch. 4 Manuscript.

Odling-Smee, F. J., K. N. Laland, and M. W. Feldman. 2003. *Niche construction: The neglected process in evolution.* Princeton University Press.

Orzack, S. H., and E. Sober. 2001. *Adaptationism and optimality.* Cambridge University Press.

Pavlicev, M., A. M. Zupan, A. Barry, S. Walters, K. M. Milano, H. J. Kliman, and G. P. Wagner. 2019. "An experimental test of the ovulatory homolog model of female orgasm." *Proceedings of the National Academy of Sciences* 116(41). 20267–73.

Pavličev, M., and G. P. Wagner. 2016. "The evolutionary origin of female orgasm." *Journal of Experimental Zoology Part B: Molecular and Developmental Evolution* 326(6): 326–37.

Pinker, S. 1999. "How the mind works." *Annals of the New York Academy of Sciences* 882: 119–27.

Puts D. A., K. Dawood, and L. L. M. Welling. 2012a. "Why women have orgasms: An evolutionary analysis." *Archives of Sexual Behavior* 41 (5): 1127–43.

Puts, D. A., L. M. Welling, R. P. Burriss, and K. Dawood. 2012b. "Men's masculinity and attractiveness predict their female partners' reported orgasm frequency and timing." *Evolution and Human Behavior* 33(1): 1–9. https://doi.org/10.1016/j.evolhumbehav.2011.03.003.

Raff, R. A. 1996. *The shape of life: Genes, development and the evolution of animal form.* University of Chicago Press.

Raff, R. A., and E. C. Raff. 2009. "Evolution in the light of embryos: Seeking the origins of novelties in ontogeny." In M. D. Laubichler and J. Maienschein (eds.) *Form and function in developmental evolutions.* 83–111. Cambridge University Press.

Reeve, H. K., and P. W. Sherman. 1993. "Adaptation and the goals of evolutionary research." *The Quarterly Review of Biology* 68(1): 1–32. https://doi.org/10.1086/417909

Reznick, D., and J. Travis 1996. "*The empirical study of adaptation in natural populations.*" In Rose, M. R., and G. V. Lauder (eds.), *Adaptation.* 243–89, Academic Press.

Rose, M. R., and V. Lauder. 1996. *Adaptation.* Academic Press. https://academic.oup.com/sysbio/article/47/3/538/1703769.

Russell, B. 1919. *Introduction to mathematical philosophy.* McMillan.

Sagan [Margulis], L. 1967. "On the origin of mitosing cells." *Journal of Theoretical Biology* 14 (3): 225–IN6. https://doi.org/10.1016/0022-5193(67)90079-3

Sansom, R. 2003. "Constraining the adaptationism debate." *Biology and Philosophy* 18 (4): 493–512. https://doi.org/10.1023/A:1025581622161

Schmitt, D. P., and J. J. Pilcher. 2004. "Evaluating evidence of psychological adaptation: How do we know one when we see one?" *Psychological Science* 15(10): 643–49. https://doi.org/10.1111/j.0956-7976.2004.00734.x

Seger, J, and J. W. Stubblefield. 1996. "Optimization and adaptation." In Rose, M. R., and G. V. Lauder (eds.) *Adaptation.* 93–102. Academic Press.

Sherman, P. W. 1989. "The clitoris debate and the levels of analysis." *Animal Behaviour* 37(4): 697–98. https://doi.org/10.1016/0003-3472(89)90052-3.

Shpak, M., and G. P. Wagner. 2000. "Asymmetry of configuration space induced by unequal crossover: implications for a mathematical theory of evolutionary innovation." *Artificial Life* 6(1): 25–43. https://doi.org/10.1162 /106454600568302.

Sinervo, Barry, and A. L. Basolo. 1996. "Testing adaptation using phenotypic manipulations." In Rose, M. R., and G. V. Lauder (eds.) *Adaptation.* 149–185.

Smith, D. 2005. "A critic takes on the logic of female orgasm." *The New York Times*, May 17, 2005, sec. Science Times. www.nytimes.com/2005/05/17/ science/a-critic-takes-on-the-logic-of-female-orgasm.html.

Smith, M. J. 1978. "Optimization theory in evolution." *Annual Review of Ecology and Systematics* 9:31–56.

Song, N., A. Lin, and X. Zhao. 2018. "Insight into higher-level phylogeny of Neuropterida: Evidence from secondary structures of mitochondrial rRNA genes and mitogenomic data." *PLOS ONE* **13** (1): e0191826. doi:10.1371/ journal.pone.0191826. ISSN 1932–6203.

Sterelny, K., and P. Kitcher. 1988. "The return of the gene." *The Journal of Philosophy*. 85(7): 339–361.

Suárez, J., and V. Triviño. 2020. "What is a hologenomic adaptation? Emergent individuality and inter-identity in multispecies systems." *Frontiers in Psychology*. 11.

Sunquist, F. 2006. "Malaysian mystery leopards." *National Wildlife*. www .nwf.org/en/Magazines/National-Wildlife/2007/Malasian-Mystery.

Symons, D. 1979. *The evolution of human sexuality*. Oxford University Press.

Symons, D. 1990. "Adaptiveness and adaptation." *Ethology and Sociobiology* 11(4): 427–44. https://doi.org/10.1016/0162-3095(90)90019-3.

Thornhill, R. 1990. "The study of adaptation." In Bekoff, M., and D. Jamieson (eds.) *Readings in Animal Cognition*. 31–62. Massachusetts Institute of Technology Press.

Tooby, J., and L. Cosmides. 1997. "Letter to the editor of the *New York Review of Books* on Stephen Jay Gould's 'Darwinian fundamentalism.'" *New York Review of Books*, June 26, 1997. http://cogweb.ucla.edu/Debate/CEP_Gould.html.

Vaccaro, C. M. 2015. "The use of magnetic resonance imaging for studying female sexual function: A review." *Clinical Anatomy* 28 (3): 324–30. https:// doi.org/10.1002/ca.22531

Van Fraassen, B. C. 1977. "The pragmatics of explanation." *American Philosophical Quarterly* 14 (2): 143–50.

Wade, M. J. 1978. "A critical review of the models of group selection." *The Quarterly Review of Biology* 53 (2): 101–14. https://doi.org/10.1086/410450

Wade, M. J. 2016. *Adaptation in metapopulations: How interaction changes evolution*. University of Chicago Press.

Wagner, G. P. 2015. "Evolutionary innovations and novelties: Let us get down to business!" *Zoologischer Anzeiger-A Journal of Comparative Zoology* 256: 75–81.

Wagner, G. P. 2000. "What is the promise of developmental evolution? Part i: Why is developmental biology necessary to explain evolutionary innovations?" *Journal of Experimental Zoology* 288 (2): 95–98.

Wake, D. B. 1991. "Homoplasy: The result of natural selection, or evidence of design limitations?" *The American Naturalist* 138 (3): 543–67. https://doi .org/10.1086/285234

Wake, D. B. 2009. "What salamanders have taught us about evolution." *Annual Review of Ecology, Evolution, and Systematics* 40 (1): 333–52. https://doi.org /10.1146/annurev.ecolsys.39.110707.173552

Wallen, K., and E. A. Lloyd. 2011. "Female sexual arousal: Genital anatomy and orgasm in intercourse." *Hormones and Behavior, Special Issue on Research on Sexual Arousal* 59(5): 780–92. https://doi.org/10.1016/j .yhbeh.2010.12.004.

Wallen, K., E. A. Lloyd, and P. Z. Myers. 2012. "Zietsch & Santtila's study is not evidence against the by-product theory of female orgasm." *Animal Behaviour* 84(5): e1–e4.

Weisberg, M. 2006. "Robustness analysis." *Philosophy of Science* 73(5): 730–42.

Wheatley, J. R., and D. A. Puts. 2015. "Evolutionary science of female orgasm." In Shackelford, T. K., and R. D. Hansen (eds.) *The evolution of sexuality*. Springer International Publishing. 123–48. https://doi.org/10.1007/978-3-319-09384-0_7

Wikelski, M., V. Carrillo, and F. Trillmich. 1997. "Energy limits to body size in a grazing reptile, the Galapagos marine iguana." *Ecology* 78 (7): 2204–17.

Wikelski, M., and C. Thom. 2000. "Marine iguanas shrink to survive El Niño." *Nature* 403 (6765): 37–38. https://doi.org/10.1038/47396

Wikimedia Commons contributors. 2018. "File:GiraffaRecurrRu.Svg." https:// Commons.Wikimedia.Org. November 29, 2018. https://commons.wikimedia .org/w/index.php?title=File:GiraffaRecurrRu.svg&oldid=278005938

Williams, G. C. 1966. *Adaptation and natural selection*. Princeton University Press.

Woodward, J. 2006. "Some varieties of robustness." *Journal of Economic Methodology* 13(2): 219–40.

Wright, L. 1973. "Functions." *The Philosophical Review* 82(2): 139–68. doi:10.2307/2183766

Wright, S. 1931. "Evolution in Mendelian populations." *Genetics* 16 (2): 97–159.

Zietsch, B. P., and P. Santtila. 2011. "Genetic analysis of orgasmic function in twins and siblings does not support the by-product theory of female orgasm." *Animal Behaviour* 82 (5): 1097–1101. https://doi.org/10.1016/j.anbehav.2011.08.002

Zietsch, B. P., and P. Santtila. 2013. "No direct relationship between human female orgasm rate and number of offspring." *Animal Behaviour* 86 (2): 253–55. https://doi.org/10.1016/j.anbehav.2013.05.011

Zuk, M. 2006. "The case of the female orgasm (review)." *Perspectives in Biology and Medicine* 49 (2): 294–98.

Acknowledgments

Thank you to Arnold and Maxine Tanis for their support of my research over many years. I owe thanks to many biologists and philosophers for discussion about the topic of this book over many years, including especially the following: The Biology Studies Reading Group at IU, Colin Allen, Linnda Caporael, Janet Collett, Michael Dietrich, Stephen Downes, Marcus Feldman, Justin Garcia, Stephen Jay Gould, Jim Griesemer, Chris Haufe, David Hull, Ryan Ketcham, Roy Levin, Richard Lewontin, Daniel Lindquist, Alan Love, Eduoard Machery, Gordon McOuat, Roberta Millstein, Elizabeth and Rudy Raff, Michael Ruse, Elliott Sober, Javier Suarez, Donald Symons, Michael Wade, Michael Weisberg, David Sloan Wilson, Stuart Newman, and two anonymous referees for *Biological Theory*. Please forgive me, those I have not mentioned due to my faulty memory!

Cambridge Elements ☰

Elements in the Philosophy of Biology

Grant Ramsey

KU Leuven

Grant Ramsey is a BOFZAP research professor at the Institute of Philosophy, KU Leuven, Belgium. His work centers on philosophical problems at the foundation of evolutionary biology. He has been awarded the Popper Prize twice for his work in this area. He also publishes in the philosophy of animal behavior, human nature and the moral emotions. He runs the Ramsey Lab (theramseylab.org), a highly collaborative research group focused on issues in the philosophy of the life sciences.

Michael Ruse

Florida State University

Michael Ruse is the Lucyle T. Werkmeister Professor of Philosophy and the Director of the Program in the History and Philosophy of Science at Florida State University. He is Professor Emeritus at the University of Guelph, in Ontario, Canada. He is a former Guggenheim fellow and Gifford lecturer. He is the author or editor of over sixty books, most recently *Darwinism as Religion: What Literature Tells Us about Evolution; On Purpose; The Problem of War: Darwinism, Christianity, and Their Battle to Understand Human Conflict; and A Meaning to Life.*

About the Series

This Cambridge Elements series provides concise and structured introductions to all of the central topics in the philosophy of biology. Contributors to the series are cutting-edge researchers who offer balanced, comprehensive coverage of multiple perspectives, while also developing new ideas and arguments from a unique viewpoint.

Cambridge Elements ☰

Philosophy of Biology

Printed in the United States
by Baker & Taylor Publisher Services

Printed in the United States
by Baker & Taylor Publisher Services